VALORIZATION IN AGRICULTURE WASTE

TO ENHANCE FOOD SECURITY

The Author

Dr. Prashant Anil Pawase, Assistant Professor at MIT ADT University, Pune, is a distinguished scholar and thought leader in the field of food technology. With an exemplary academic trajectory, Dr. Pawase holds a B.Tech, M.Tech, and Ph.D. in Food Technology from premier Indian institutions, where he consistently demonstrated academic excellence. His remarkable achievements include securing top ranks in the highly competitive ICAR and MCAER entrance examinations, reflecting his dedication and aptitude for agricultural and food sciences. Dr. Pawase has an impressive record of scholarly contributions, including 33 peer-reviewed publications in esteemed journals and presentations at numerous national and international conferences. His innovative spirit is evidenced by his ownership of three patents in food technology, showcasing his commitment to advancing the field through practical and impactful research. A testament to his academic rigor, he is a four-time qualifier of the prestigious Agricultural Scientists Recruitment Board (ASRB) NET and has successfully cleared the Agricultural Research Services (ARS) mains and interviews, further highlighting his expertise and leadership in agricultural research and education

VALORIZATION IN AGRICULTURE WASTE

TO ENHANCE FOOD SECURITY

Prashant Anil Pawase

2026

Daya Publishing House®

A Division of

Astral International Pvt. Ltd.

New Delhi – 110 002

ISBN: 978-93-5919-298-7

Published by : **Daya Publishing House**®
A Division of
Astral International Pvt. Ltd.
– ISO 9001:2015 Certified Company –
4736/23, Ansari Road, Darya Ganj
New Delhi-110 002
Ph. 011-43549197, 8130496929
E-mail: info@astralint.com
Website: www.astralint.com

Printed at : **Replika Press Pvt. Ltd.**

This book is dedicated to all the farmers, scientists, and innovators who labor day and night to feed the world and protect our planet. Their grit, ingenuity, and passion for sustainability are the forces driving the valorisation of agricultural waste. To the farmers who labor under the sun and rain, cultivating the soil that feeds us all—your labor is the backbone of our food system. To the innovators and researchers who can spot potential in the waste seen by others—your creativity illuminates the way towards a green tomorrow. And to generations to come—may this work be one step toward a world where nobody goes without, and the riches of our world are nurtured and wisely used. Let us together make waste into wealth and scarcity into abundance.

PREFACE

In a world struggling with the combined challenges of population increase and environmental degradation, the search for sustainable options to improve food security has never been more critical. The key to human existence, agriculture, produces tons of waste, which in many cases go to waste, polluting the environment and wasting resources. But in this challenge is an unbelievable chance—the valorization of farm waste. By converting what is generally regarded as "waste" into useful resources, we are able to meet essential challenges like food shortages, malnutrition, and environmental sustainability. This volume, Valorization of Agricultural Waste to Enhance Food Security, is the fruition of research, creativity, and on-the-ground experience that seeks to unlock the value of agricultural by-products. Innovative technologies, green practices, and policy frameworks are mapped here that can transform the possibility of agricultural waste into a pillar of food security. From biofertilizers and animal feed to bioenergy and nutrient-dense foods, the choices range as widely as they are compelling.

The path to a food-secure world involves working together across disciplines, sectors, and borders. This book is an invitation to researchers, policymakers, farmers, and industry leaders to seize the overlooked value in agricultural waste. It is my hope that the solutions and ideas proposed here will inspire and empower readers to help build a more equitable and sustainable food system.

CONTENTS

ABBREVIATIONS

AA- Antioxidant activity
ABE- Acetone-butanol-ethanol
ABTS- 2,2′-azino-bis-(3-ethylbenzothiazoline-6 sulfonic acid)
BA- Bioavailability
BC- Bioactive compounds
BCBA- β-carotene bleaching antioxidant capacities
BEE- Betalain encapsulation efficiency
BI- Bioaccessibility
BWE- Beetroot waste extract powder
CCPE- Commercial citrus peel extracts
CSE- Conventional solvent extraction
DPPH- 2,2-diphenyl-1-picrylhydrazyl
DW- Dry weight
EA- Encapsulating agents
EAE- Enzyme-assisted extraction
EE- Encapsulation efficiency
FD- Freeze drying
FF- Functional food
FRAP- Ferric reducing antioxidant power
FV- Fruits and vegetables
FVW- Fruit and vegetable waste
GA- Gum arabic
GEL- Gelatin
HMS- High-maize starch
HSH- High-speed homogenizer
HWPI- Hydrolyzed whey protein isolate
IN- Inulin
IT- Inlet temperature
MAE- Microwave-assisted extraction

MD- Maltodextrin

MEC- Microwave extraction of chlorophyll

MPE- Mandarin peel extract

MSK- Mango seed kernels

MSKE- Mango seed kernel extract

NHWPI- Non-hydrolyzed whey protein isolate

OSWE- Onion skin water extracts

OT- Outlet temperature

PCEP- Polyphenolic compounds extract powder

PEFE- Pulsed electric field-assisted extraction

PLE- Pressurized liquid extraction

ROS- Reactive oxygen species

SA- Sodium alginate

SC-CO2- Supercritical CO2 extraction

SD- Spray drying

SFE- Supercritical fluid extraction

SPI- Soy protein isolate

ST- Soluble starch

SWE- Subcritical water extraction

TC- Total carotenoids

TE- Trolox equivalents

TEC- Traditional extraction of chlorophyll

TF- Total flavonoids

TPC- Total phenolic content

UAE- Ultrasound-assisted extraction

WP- Whey protein

WPC- Whey protein concentrate

WPI- Whey protein isolate

WRP- Watermelon rind powder

CHAPTER 1

INTRODUCTION OF AGRICULTURAL WASTE AND FOOD SECURITY

Overview of Agricultural Waste

Definition and types of agricultural waste (crop residues, animal waste, processing by products) The global scale of agricultural waste generation

The amount of agricultural trash produced worldwide Agricultural materials are those that are created on Earth when the climate changes. They are essentially the organic materials that are consumed by both people and animals. Following processing, these organic ingredients are turned into agricultural waste.

What is the agriculture waste??

Agricultural waste is the term used to describe byproducts of farming operations, including crop remnants, livestock excrement, and residual plant components. It contains a wide variety of materials, both organic and inorganic. Due to its high conversion potential, biomass is now produced from agricultural waste in order to produce electricity. Biomass also acts as a contingency energy supply. One effective method of converting agricultural waste into energy is to use it to create biomass (Harshwardhan and Upadhyaya, 2017).

One of the biggest economic sectors is agriculture, which produces a vast amount of waste from different agricultural activities. Globally, about 140 billion metric tons of biomass from agricultural waste are produced. The necessity for increased agricultural production was prompted by the human population explosion, which ultimately led to the development of enormous amounts of agricultural waste. Over the past ten years, developing nations like China and

India have seen fast economic growth and population expansion, which has increased trash generation. India generates around 350/990 million tons of solid trash annually, making it the most prolific country in Asia (Pappu et al., 2010).

Approximately 620 million tons of agricultural waste are produced annually in India, an agro-based nation (Patel et al.,2020). The availability of biomass from agricultural waste in large quantities has made using these wastes as an alternative energy source essential. (Harshwardhan *et al.*, 2008). Additionally, human health has suffered greatly as a result of intensive farming techniques (Muhammad et al., 2020). Serious environmental problems arise from improper management techniques such as the disposal of agricultural waste and the open burning of bare land to remove the garbage (Demirbas *et al.*, 2011).

Additionally, these activities produce a large amount of harmful pollutants (N2O, SO2, CH4, and smoke) that have an impact on the atmosphere globally (Pardo *et al.*, 2017). There are two primary categories of agricultural wastes and food and agro-industry products used as animal feed:

1. Plant origin
2. The fermentation sector

The residues from the food and agriculture industries account for almost 30% of the world's agricultural output. The majority of these wastes are made up of lignocellulosic materials, waste from fruits and vegetables, waste from the sugar sector, and trash and byproducts from animals and fisheries. Agro-residues are rich in a variety of bioactive and nutraceutical compounds, such as dietary fiber, carotenoids, and polyphenolics. Agricultural residues, a substantial and valuable biomass, offer prospective solutions to problems pertaining to animal nutrition and the world's supply of calories and protein, provided that appropriate techniques are used for their value-adding through nutrient enrichment.

Technologies for protein enrichment of these wastes include installation, high solid or slurry processes, and solid substrate fermentation. When creating technologies for recycling garbage, it is important to take into account the distinct qualities of each waste as well as the environment in which it is generated, processed, and consumed. More precisely, these technologies need to generate safe products for both animal feed and human consumption. This review focuses on the primary current applications of solid-state fermentation in the feed business (Ajila *et al.*, 2012). Crop residues are a certain by-product of crop harvest, and their quantity is directly related to all the elements that typically affect crop output. However, research and development is heavily focused on the quality of these residues as an easily accessible source of animal nutrition. This is owing to the fact that the majority of crop wastes are high in fiber and low in nutrients. Therefore, giving these residues to animals without treatment may hinder their acceptance and, in turn, their performance. Therefore, these residues may satisfy the majority of the animals' dietary needs at comparatively low costs if the quality and amount of feedstuff were appropriately correlated with the nutrient requirements of the species or breed of animals. Together, this are accomplished by means of procedures like milling and enrichment initiatives such

Crop residue

The fibrous plant components left over after a crop is harvested for human consumption are known as crop leftovers. In addition to crop harvest, cereal crops yield enormous amounts of stem and leaf, which are then fed to livestock and turned into profitable products. Proper storage makes the most of these wastes, ensuring that the animals have a sufficient feed during the lean season when fodder is in short supply. The various sources of crop residues are as shown in the table below;

1. Straw

The stems and leaves of tiny grains that remain after the crop is harvested are known as straw, and they typically make up more than half of the crop's harvestable vegetation. In most societies across the world, straw makes a decent maintenance diet for ruminants. However, because of its weak nutritional makeup, it is not often considered a full ration for production animals; instead, it is balanced with high-quality green fodder and concentrates to provide a total mixed ration. The primary sources of straw harvest are rice and wheat, with smaller amounts coming from barley, millet, oats, and practically every other cereal crop that yields high-quality and tasty roughages. These straws are simple to gather, handle, and bale using either traditional methods using standard haymaking equipment or automated systems (Fertahi *et al.*, 2023)

2. Rice straw

The vegetative portion of the rice plant, known as rice straw, is often chopped during or after grain harvest. Rice straw is unique among grains since the stem is easier to digest than the leaves. Therefore, cutting it as near to the ground as feasible is regarded as economically advantageous. Traditional harvesting produces higher-quality straw than that from mature plants since the crop is frequently collected while the straw is still relatively green. Almost all rice-growing nations frequently feed rice straw to ruminants, particularly mature draught stock or even productive animals. Green fodder and concentrate supplements are typically used while feeding rice straw (Dobermann *et al.*,2002).

3. Wheat straws

Wheat straws are important source of nutrition in traditional wheat-growing countries of Asia and Africa, but in most parts of Europe it is considered very poor, and not fed except in emergency, usually after ammonia treatment. Harvesting and processing technology is similar to the rice straw, but developments in harvesting methods have effects on straw quality to some extent. With the combine harvesters and grain drying, the crop is cut at a slightly mature stage, which further causes more leaf loss and straw of lower feeding value. Wheat straws which also also help to reduce green house gas emission (Zhang *et al.*, 2025).

4. Stubble

The stumps of harvested crops, such as grains and legumes, that are left on the field after harvest are known as stubble. Small-scale farmers in several developing Asian and African nations frequently graze their stubble. Along with the base of the crop harvest, stubble frequently contains important feed in the form of weeds and grain that was overlooked during the harvesting process, particularly when harvesting by machine. The livestock benefit from stubble grazing because it enables them to graze selectively and likely find at least a maintenance diet from what, when considered as a whole, could be a sub-maintenance feed. On the other hand, trampling from on-field grazing results in significant waste (Abdurrahman *et al.*, 2020).

5. Stover

Stover is made up of the leftovers from growing big crops like sorghum and maize. In many regions of the world, these remnants represent important forages. Stover is typically handled and dried in its long, uncut state in small-scale production systems, frequently by standing in or on the field's edge before being stored. It is also baled and ensiled, either with or without urea treatment, in large-scale systems (Gupta *et al.*, 2022).

6. Corn Stover

The complete plant remnants that remain in the field following corn harvest are known as maize stover. It consists of cobs, husks, leaves, and stalks. Corn stover is inexpensive, widely available, and has a lot of room to grow and be used. In order to save money on winter calf feed, some small-scale farms harvest and dry the stover in the field or at the farmhouse after the corn harvest, while others graze it in the field. On larger farms, it is either ensiled or baled after drying. Compared to most straws, maize stover contains more nutrients. Sweet corn cobs, which are harvested while the plant is still green, are one of the corn kinds that provide a significant amount of high-quality roughage as a by-product. A few days after the cobs are harvested, this stover is left to grow. When compared to stover from a fully-mature crop, the stover from such plants provides a better feed. According to the studies, the nutritious content of various maize plant sections varies greatly. Although they only make up around 12% of the residue left in a field, the husks are more nutrient-dense and easier to digest than leaves. The mature, non-lactating cow can be sustained by the leaves (Kim *et al.*, 2009).

7. Sorghum stover

One of the cereal crops that grows best in hot, arid regions of Asia, Africa, and portions of America and Australia is sorghum. The stover is harvested as soon as the crops are harvested, ideally by cutting it as green as possible so that it can be dried and used as corn for future feeding. In certain nations, grazing is practiced, but it is not ideal due to the possibility of waste from manure and trampling, which is always the case when Stover is grazed. From sorghum stover which produce biofuels for reduce waste (Stamenkovic *et al.*, 2020)

8. Sugar cane by products

In order to produce sugar, charcoal, alcohol, and biofuel, sugar cane is collected. Cane tops, terminal leaves, nodes, and bundle sheath are among the on-farm by-products of cane harvesting that are used to make animal feed. Additionally, cane yields other industrial by-products including bagasse and molasses, but the latter requires treatment for the remaining sugar in the pressed stalk because of its high content of poor fermentable fiber, which has a detrimental influence on the voluntary feed intake of the entire diet. The cutting point is typically at the highest fully formed node, though it might vary depending on the needs of the millers. Cane tops can be ensiled, chaffed and given directly, or separated using basic crushers, with the juice used for monogastrics and the residue for ruminant feed. Although the tops are easily assimilated by animals, they are low in protein and need to be supplemented. Final molasses, which is produced at a rate of roughly 3 tons per 100 tons of stalk entering the facility, is the most significant by-product of factory sugar production. The soluble residue left over after the sucrose in cane juice is extracted is called molasses. In addition to minerals, organic acids, and other nutrients, it offers a rich source of easily digested energy. Molasses and sugarcane juice both contain the fermentable carbohydrates fructose, glucose, and sucrose (Rajaeifar *et al.*, 2019).

9. Vegetable waste/residue

Cereals, which generate a lot of residues, are the greatest crop in the human food chain, followed by vegetables. Harvesting cabbage yields up to 6 tons of edible dry matter per hectare, making it one of the major sources of vegetable leftovers. The food can subsequently be used as waste crop residue for animal feed when vegetable markets, like those for potatoes, collapse because it is too costly to take it to market. A decent ruminant feed consists of carrot tops as well as carrots that were damaged during harvest or thrown out due to their low quality. Broccoli, cauliflower, and radish tops are examples of brasicca vegetable wastes that have similarly significant nutritious components and can be fed either fresh or dried (Bakshi *et al.*, 2016).

10. Residues of pulse and legume crops

Some commonly farmed pulses, like soybean, green and black grams, and the leafy sections of pigeon peas, can also be fed with their leaves and stems. A useful by-product, green vines from pea haulms are best preserved as silage. Hand-harvested soybean haulms and pea straw are good sources of roughage. Although they are far more difficult to recover, several of these have a higher feeding value than cereal straws. The leaves often break in dry conditions and discolor or drop at or before harvest in humid areas. It is simpler to recover the leaves and stems during threshing when the crop is dried at the homestead. However, favorable climatic conditions are crucial for successful harvesting, and by harvest time, many leaves are senescent or infected (Bhandari, B. 2019).

11. Soybean forage

The soybean is a leguminous plant that can reach a height of one meter. About 20% of soybeans are oil, making them predominantly an oilseed. The primary residues from soybeans comprise oil, which is harvested as cakes, and grain, which is harvested as stems, leaves, and hulls. These materials make up the fodder portion, which can be grazed, ensiled, or dried to create hay. In addition to having a high nutritional content and exceptional digestion, the leaf is highly appetizing to cattle. Dry cows and heifers can be fed the stubble by grazing or chopping it. The leftover material from threshing beans, known as soybean straw, can be fed to cattle as roughage (Nielsen, 2011).

12. Sunflower by products

The main purpose of sunflower cultivation is to produce oil, and the plant's remains, including its heads and remaining stalks, are sometimes used as animal feed. When there is a lack of roughage or drought, which reduces the possibility for seed production, sunflowers make a useful alternative. Typically, sunflower fodder is provided as silage. If the sunflower heads have been dried before threshing, which is highly dependent on the climate, they provide a great feed. Sunflower screens can be created by combining the dried heads with other leftovers from sunflower harvest and oil extraction, or they can be turned into a meal (Yegorov *et al.*, 2019).

13. Groundnut tops

Very useful hay is made from groundnut haulm in regions with favorable drying conditions during harvest. When the leaves start to turn yellow and the haulms start to dry out, groundnuts are ready to be harvested. Groundnut tops are carefully collected and then dried at the farmhouse, on house roofs, and in other sunny locations away from cattle in the drier regions of Asia. Another option is to stack them on stakes, let them dry for three to six weeks, and then thresh them. Animals are fed the dried straw (Desmae and Sones., 2017).

14. Miscellaneous residues

Not all agricultural leftover that can be fed to animals is dry. Though they are saved by ensiling rather than drying, sugar beet tops provide a good feed. Like good beans, sweet potato tips make a great meal. In the yam belt of West Africa, backyard sheep and goats are traditionally fed yam peelings. Although they can be ensiled, banana pseudo-stems are a valuable roughage that is frequently utilized as cattle fodder. Although they are typically only sometimes accessible, some tree trimmings make good nutrition, particularly for small ruminants. When the olive is pruned, it yields a lot of twigs and leaves, which make it a very tasty meal for animals. It is also possible to preserve low-quality fruit and part of the waste from processing fruits and vegetables for use as animal feed (Fahmy *et al.*, 2017).

In agricultural waste from animal waste produce

1. Manure (solid and liquid)
2. Urine
3. Bedding materials (e.g., straw, sawdust)
4. Dead animals

    (Girotto, F and Cossu, R. 2017)

Understanding Food Security

Security of food the concept that everyone always has access to enough safe, nourishing food to live active, healthy lifestyles, including physical, social, and financial access (FAO 2009).

Definition and dimensions of food security (availability, access, utilization, and stability) Problems about food security were initially raised during the 1948 Universal Declaration of Human Rights, which acknowledged the right to food as a fundamental component of a sufficient standard of living. Following the global oil emergency and the ensuing shortage of food in 1972–1974, public concerns about national and international food security quickly increased. Since the World Food Conference in 1974, about 200 definitions of food security have been put forth. The first definition of food security was the availability (at all times) of sufficient global supplies of staple foods to support a continuous increase in food consumption and to counteract changes in prices and production1. Concerns about the quantity and stability of food supplies around the world were reflected in this definition. (FAO 2009).

In the latter part of the 1980s, food security obtained importance, and it remained the main topic of discussion throughout the 1990s. Accessibility was added to the definition in 1983, meaning that people should always have both financial and physical access to the food they require. The temporal dynamics of food insecurity—that is, temporary and chronic food insecurity—were later the focus of the World Bank. While ongoing hunger is linked to issues that remain because of insecurity and low incomes, transitory food insecurity involves times of increased pressure that may be brought on by natural disasters or economic collapse. Then, food security is further defined as ensuring that everyone always has access to enough food for an active (Frankenberger T *et.al* 1995)

There are numerous definitions of food security, which can be used to identify fundamental concepts. According to Maxwell and Frankenberger, the concept of food security is defined by a number of definitions, and the main feature that differentiates household food security is "secure access at all time to enough food," which is derived from the concepts of secure, access, time, and enough. From the individual and home level (micro level) to the community level (meso level), and finally to the national and international level (macro level), the notions of food security availability, accessibility, acceptability, and sufficiency apply to all levels of social and administrative institutions. (Weingartner 2000). The number of hungry people will increase if current trends continue as predicted. Previous efforts to address the short-term, acute needs of hungry people will probably impact future conversations about how to handle food shortages and other food emergencies. since of this need, talks about food security frequently center on abrupt and painful disruptions like famines, wars, and natural catastrophes since these events garner a lot of attention from international organizations, the media, and governments.

Dimensions of Food Security

The 'four pillars' of food security are generally understood to be (1) availability, (2) access, (3) usage, and (3) stability (FAO, 2006; Webb et al., 2006; CFS, 2009; Upton et al., 2016). This four-pillar approach only gradually emerged over the course of 50 years, despite the fact that we tend to think of these pillars of food security as obvious and second nature when we use the term today, especially because they give us a shorthand summary of some of the key components of the concept. In actuality, these varied elements rose to prominence at different points in history as a direct result of broader occurrences and academic discoveries that led to a greater comprehension of the various facets of food security and its importance in governance. Even though the aspects following availability were not included in the previously mentioned sequence, the four pillars availability, access, usage, and stability are nearly invariably mentioned in that order today. The Integrated Food Security Phase Classification (IPC) guidelines, for instance, make reference to the pillars and emphasize that field assessors should believe that these pillars interact sequentially: "food must be available; households must have access to it; they must utilize it appropriately; and the whole system must be stable" (IPC, 2019)

1. Availability

Taking availability into account, there are two ways to supply food: either domestic production or imports. The actual availability of food at farms and local markets must be carefully considered in light of this. This in turn depends on market infrastructures that are in good working order, including sufficient road and rail networks, as well as the provision of sufficient technology for processing and storage (Riely, F. *et al* 1999)

2. Access

Ensuring that individuals have sufficient physical and financial access to food through growing it, buying it, receiving it as a gift, bartering or trading for it, and other means is known as food access (Riely, F *et al* 1999). This idea can be viewed as a set of rights that enables a person to obtain and preserve the right foods for a sufficient diet and level of nutrition. This could be acquired directly, as described by growing your own food, making enough money, bartering, and exchanging it, or indirectly, through social arrangements at the local or national level, such as family, welfare systems, traditional rights, access to shared resources, and, of course, emergency food assistance (Sen, A. Poverty and Famines 1981).

3. Stability

The understanding that food security can be lost as well as gained is a growing issue in the food security discussion, even though it is not a novel concept [United States Agency for International Development]. Consequently, the concept of risk management is becoming increasingly accepted as a means of combating hunger. [United States Agency for International Development] Such consideration involves issues of stability and vulnerability, whether they are related to livelihoods in particular, the larger economy in general, incomes, or even food supplies themselves, with a focus on sudden or other shocks like floods, droughts, or pests, among others.

4. Utilisation

The final basis is biological utilization, which is not a unique concept despite the "novelty" of the terminology. Biological utilisation is simply the ability of an individual to absorb food as best they can, or as effectively as possible. Research has also revealed a strong correlation between this skill and an individual's health state, which is based on significant non-food inputs. In fact, it has been demonstrated that optimal biological use requires the provision of clean water and sanitation services, appropriate health and child care, and sufficient knowledge of physiological and nutritional requirements, as well as the appropriate application of that knowledge.

Linking agricultural waste and food security The potential of agricultural waste in addressing food security

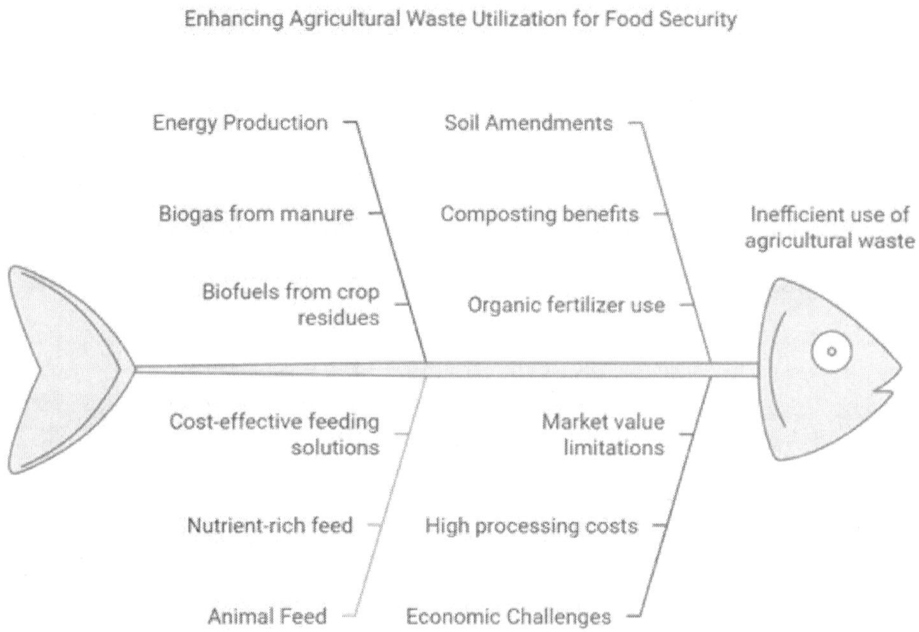

Making agricultural waste into a resource that can be used rather than simply thrown away should be the goal. The primary ways that agricultural wastes can improve food security are by producing energy, using them as animal feed, and using them as soil amendments and biofertilizer. Many of them are safe to add directly to the soil and contain significant amounts of organic matter. (Sabiiti, E. N. 2011)

One of the waste treatment technologies that enables the use of organic waste as fertilizer, even in populated areas, is the conversion of these agricultural wastes (crop residues and animal manures) into organic fertilizers (via composting). In order to improve soil fertility and, consequently, agricultural output, technology is essential. (Sabiiti. et.al., 2018) In most of Africa, where a lack of nutrients poses a significant barrier to food production, the use of organic fertilizers is especially crucial. Additionally, composting eliminates pathogens, lowers weed germination in agricultural fields, eliminates odors, and reduces the volume of garbage, all of which help to address major environmental issues related to the disposal of huge amounts of waste. The compost can be utilized on the same farm or sold to generate extra income. Additionally, the growing demand for organic items like maize and goat meats has led to an increase in the manufacture of composts for agricultural usage. (Sanchez. *et. al.* 1997)

The remnants from the cultivation and processing of raw agricultural products, including fruits, vegetables, meat, poultry, dairy products, and crops, are referred to as agricultural wastes. These are the byproducts of agricultural production and processing that may contain materials that are useful to humans but whose economic worth is less than the expenses of gathering, transporting, and processing them for useful purposes. They might be liquids, slurries, or solids, and their composition will vary depending on the system and kind of agricultural activity. (Khaleel *et al.*, 1981) Animal waste (manure, animal carcasses), food processing waste (only 20% of maize is canned, and 80% is waste), crop waste (corn stalks, sugarcane bagasse, fruit and vegetable drops and culls, prunings), and hazardous and toxic agricultural waste (pesticides, insecticides, herbicides, etc.) are all considered agricultural waste, or agro-waste. Although there are few estimates of the amount of trash generated by agriculture, it is generally believed that this sector accounts for a sizable amount of all garbage in the developed world. Naturally, as agricultural production has expanded, so too have the amounts of livestock manure, crop wastes, and agro-industrial byproducts. (Agamuthu, P. 2019)

The most recent individual valorization technologies designed to produce value-added byproducts from food wastes, such as bio-based adsorbents, bio-fuels, bio-materials, and value-added components. The objective is to evaluate the process parameters influencing the effectiveness of the production of the bio-based products and to look at the related benefits and limitations of each technique independently. The difficulties encountered in converting the waste into each of the bioproducts have been described, along with the potential applications in the future (Nayak and Bhushan., 2019). The World Food Programme (WFP), the biggest hunger-fighting humanitarian organization in the world and the main source of information about food insecurity in sub-Saharan Africa, provides a unique chance to study the modern relationship between food and security. This article looks at the WFP's "turn" toward resilience in its procedures and guidelines. Our research highlights that the WFP aims to regulate food security through a variety of security methods, of which resilience is one.

Therefore, it cannot be said that resilience is replacing security as the primary rationale driving insecurity, as some have claimed. Nonetheless, a key component of the WFPs' ongoing work is resilience. In the context of food crises, resilience is a way of thinking that acknowledges the unavoidability of unpredictable, high-impact events and seeks to promote the capacity for systems and people to adapt, absorb, and recover from their effects. This is in contrast to more well-known security strategies that try to prevent or contain disruptive events. Although resilience is promoted as a component of a comprehensive approach to address a variety of issues facing human populations today, its sole goal is to give individuals and groups the ability to cope with the complexities of a neoliberal food system without challenging, upending, or opposing the fundamental causes of socioeconomic and political instability (O'Connoretal.,2017).

Nearly one-third of all food production is wasted as food loss and waste (FLW), despite the fact that food shortages are still a major issue in many parts of the world. Despite the fact that global food production is anticipated to be constrained by environmental and resource limits, this is linked to around 25% of the land, water, and fertilizer required for crop cultivation. FLW reduction has lately drawn a lot of interest since it offers a chance to improve environmental sustainability and food security. This study emphasizes the significance of FLW prevention as an additional strategy to solve the Grand Challenge of global food security and environmental sustainability by examining recent advancements and novel breakthroughs in the literature. However, achieving the anticipated FLW reduction requires more than just increasing awareness. By combining the FLW reduction techniques and the obstacles, such as

1. completing the data gaps
2. calculating the socioeconomic and environmental effects of FLW reduction strategies, we are able to identify the knowledge gaps and research possibilities.
3. being aware of the impacts of size, and
4. investigating the effects of worldwide shifts.

More aggressive yet scientifically grounded measures are urgently needed to minimize FLW, and they call for the participation of all parties involved in the food supply chain, including policymakers, food producers and suppliers, and food consumers (Shafiee-Jood & Cai 2016).

A result of farming and animal husbandry is agricultural waste (Gontard et al., 2018). Unharvested crops and animal excrement are examples of agricultural waste. Biomass, the organic material found in these wastes, can be used to produce animal feed, biofertilizers, and bioenergy (Haque *et al.*, 2023; Khan *et al.*, 2023).

As many farmers have done for years, there are straightforward ways to use agricultural waste, such as turning it straight into animal feed or firewood (Haque *et al.*, 2023; Khan *et al.*, 2023). Agricultural waste can also be recycled in a number of ways to produce improved goods. The procedures include extraction (to produce biodiesel), biochemical (to produce bioethanol and biogas for fuel), and thermochemical (e.g., to produce heat for power) (Khan *et al.*, 2023). The general use of agricultural waste is seen in Fig.

Making the most of agricultural waste utilization could significantly improve economic benefits and preserve the environment (Haque et al., 2023; Khan et al., 2023). However, there are still several barriers to the efficient use of agricultural waste in developing nations. These challenges include farmers' limited awareness of the environmental hazards posed by agricultural waste, their ignorance of technological developments and production techniques for products derived from agricultural waste, and the lack of incentives brought on by the low demand for these products (Haque *et al.*, 2023)

CHAPTER 2

AGRICULTURAL WASTE – SOURCES AND TYPES

Crop Residues, Straw, stalks, husks, and other field residue, Nutritional and industrial potential of crop residues

Introduction

After crops have been harvested and threshed, crop residues refer to the leftover parts of the plants in the field. Recycling crop residues is advantageous as it converts farm waste into a useful product to meet the nutrient requirements of subsequent crops. India, being an agrarian country, produces a significant amount of agricultural waste, which is expected to increase due to the growing population and the need for improved productivity. Agricultural residues, such as cereal straws, woody stalks, and sugarcane leaves/tops, are generated annually after the harvest, along with a large amount of residues produced during milling of farm produce. These residues are utilized as animal feed, thatching for rural homes, residential cooking fuel, and industrial fuel (Niveta *et al.* 2014).

Most farmers use wheat straw to feed animals, while managing rice straw poses a challenge due to its high silica content, making it a poor animal feed. The leftover rice residues from the combine harvester interfere with the planting of wheat using a seed drill, leading farmers to burn the rice straw (90-140 Mt annually). Burning is considered the most suitable method for farmers, as it is cost-effective and acts as an effective pest control procedure (Adam John 2013).

The burning of crop stubble leads to air pollution and the loss of significant biomass, including organic carbon and plant nutrients. Approximately 80–90% of carbon, 25% of nitrogen, 20% of phosphorus, 50% of sulfur, and 25% of potassium present in crop residues are lost in the form of gaseous and particulate matters, causing atmospheric pollution and global warming, as well as adverse effects on soil properties, flora, and fauna (Jat, M. L., *et al.* 2014 and Mehta, S., *et al.* 2018).

The increased mechanization, particularly the use of combine harvesters in the IGP, and the decline in traditional use of crop residues for domestic purposes, are major causes leading to forced residue burning by farmers. Crop residue is typically considered waste, but if properly utilized and recycled, it can play a crucial role in fulfilling nutrient demands in the cropping system (Kumar *et al.* 2015). Around one-fourth of nitrogen and phosphorus, three-fourths of potassium, and half of sulfur uptake by cereal crops are retained in crop residues, making them valuable organic nutrient sources. This can help bridge the gap between the recommended nutrient doses and what farmers are actually following, which is lower in major cropping systems (29% lower level of nitrogen, 25% lower level of phosphorus, 71% lower level of potassium, and 100% lower level of micronutrients) (Singh *et al.* 2017a, b, c). According to Sidhu *et al.* (2007), paddy straw contains 39 kg/ha of nitrogen, 6 kg/ha of phosphorus, 140 kg/ha of potassium, and 11 kg/ha of sulfur. Although an exact figure for the potential crop residue generation in the country is not available, different researchers have estimated it based on the area under cultivation, crop productivity, and straw-to-grain ratio (SGR), residue-to-produce ratio (RPR), or crop-to-residue ratio (CRR) (Gadde *et al.* 2009; Hiloidhari *et al.* 2014; Lohan *et al.* 2017). According to various estimates by researchers, the crop residue potential of India is 620 million tones (MT) (Sharma, A., *et al.* 2020).

Crop Residue Management

Management of crop residues in the rice-wheat cropping system in India leads to the production of large quantities of crop residues, especially with the majority of rice and wheat harvesting in North West India being done using combine harvesters, leaving residues in the field. The residues of cereal crops, mainly rice straw and husk, serve various purposes such as cattle feed, domestic fuel, mulching, and use in boilers for parboiling rice. However, the management of rice straw presents a serious challenge due to the limited time between rice harvest and wheat sowing, as well as the higher silica content compared to other crops. Farmers have various management options available for crop residues, including livestock feed, mushroom cultivation, incorporation, surface retention and mulching, as well as removing the straw, and they choose different straw management practices based on their specific situations (Singh, R. V., *et al.* 2021).

Methodology

Various disciplines have conducted studies on FBCR, but these have been isolated or limited to related aspects. Efforts have been made to compile relevant information on the overall picture of the rice-wheat cropping system, health and environmental issues, farmers' concerns, and the current efforts of the Indian government and other stakeholders to stop residue burning in the IGP. Current information regarding farmers' perspectives and government initiatives has been collected from leading e-newspapers and confirmed through surveying and interacting with progressive farmers and researchers involved in field experiments/demonstrations (Patel, R., & Kumar, P. 2021).

1. Status of crop residue in India.
2. Extent of residue burning in IGP
3. Impact of residue burning on human health
4. Environmental issues associated with residue burning
5. Why farmers go for field burning of crop residues?
6. Why farmers do not feed rice straw to their animals?
7. Government initiatives
8. Residue management options:-
   - On-farm residue management
   - Rice straw as animal feed
   - Industrial utilization
9. Suggestions and recommendations for future strategies (Raghuveer Singh, *et al.*2018),

Crop Residue Management (CRM)

Crop residues are significant sources of plant nutrients and play a crucial role in the stability of agricultural ecosystems, with about 400 million tons of crop residues produced in India alone. The double cropping system heavily depletes the soil of its nutrient content, as both rice and wheat are exhaustive feeders. Management of crop residues should address global warming and food security concerns by increasing carbon sequestration in agriculture and boosting grain yields. In the North China Plain, different crop residue management practices have been historically experienced, with direct burning in the field remaining an environmental problem, while crop residue amendment triggers beneficial cycling of carbon and nutrients in agriculture (Singh, Y., & Sharma, P. 2020).

CRM systems encompass conservation tillage methods like zero-till, reduced till, bed planting, and other techniques that ensure adequate residue cover to shield the soil surface from erosion caused by wind and rain. CRM functions throughout the year, incorporating all field operations that impact residue quantity, its alignment with the soil surface and prevailing wind and rainfall patterns, and the uniformity of residue distribution during the protection period. Crop residues and their appropriate management play a role in weed suppression through physical presence on the soil surface as mulch, limiting solar radiation penetration beneath the mulch layer, direct suppression, and regulation of nitrogen availability. CRM and tillage practices also affect the efficacy of soil-applied pre-emergence herbicides. The quantity and quality of residues, as well as ash resulting from residue burning, can impact their effectiveness, often necessitating a higher herbicide application rate for effective weed control. The primary benefits of CRM compared to conventional systems include savings in fuel and labor, as well as long-term improvements in soil structure and fertility (Powlson, D. S., et al. 2014).

CHAPTER 3

TRADITIONAL USES OF AGRICULTURE WASTE

Agricultural residues are the unused byproducts of producing and processing agricultural goods that may contain useful materials but are not economically valuable enough to justify the cost of collecting, transporting, and processing for beneficial use. There are few estimates available for the amount of agricultural waste generated, but it is generally believed to contribute a significant portion of total waste in developed countries. The advancement of agriculture often leads to waste due to the excessive use of intensive farming methods and chemicals, which has a noticeable impact on rural and global environments. Agricultural waste is typically produced from various sources including farming, livestock, and aquaculture. These wastes are currently utilized for various purposes through the 'reduce, reuse, and recycle' waste management strategy. The agricultural waste management system (AWMS) and a standard waste management approach for a poultry farm were discussed using the six functions of agricultural waste management. Agricultural waste has the potential to be harmful to plants, animals, and humans through direct and indirect means. The environmental impacts of these toxic agricultural wastes were also examined, along with their management. The residues from the cultivation and processing of raw agricultural items like fruits, vegetables, meat, poultry, dairy products, and crops are known as agricultural wastes. These are the by-products of agricultural production and processing that may contain materials beneficial to humans, but their economic value is not sufficient to cover the costs of collection, transportation, and processing for beneficial use. Their composition is influenced by the agricultural system and activities, and they can exist in the form of liquids, slurries, or solids (Nwakaire *et al.*, 2016).

According to Sharma *et al.*, (2019), Agricultural waste, also known as agro-waste, consists of animal waste such as manure and animal carcasses, as well as food processing waste, with only 20% of maize being canned and the remaining

80% considered waste. Additionally, it includes crop waste like corn stalks, sugarcane bagasse, drops and culls from fruits and vegetables, and prunings, along with hazardous and toxic agricultural waste like pesticides, insecticides, and herbicides. Although estimates of agricultural waste are scarce, it is generally believed to contribute a significant portion of the total waste in the developed world. The expansion of agricultural production has naturally led to increased amounts of livestock waste, agricultural crop residues, and agro-industrial by-products. If developing countries continue to intensify farming systems, there is likely to be a substantial increase in agricultural wastes globally.

Historical Perspectives on Agriculture Waste Utilization

Using agricultural waste is an age-old practice. History is witness to the fact that resource constraints made people creative enough. Early agricultural communities relied heavily on resources such as crop residues, animal products, and other forms of farm waste. This essay outlines the history of using agricultural wastes and their developments, starting from the use of wastes in the traditional practices and ending in the present modern practices with special emphasis on important developments and their bearing on agriculture and sustainability.

Agricultural waste is material that is surplus after the cultivation and processing the agricultural products and these include fruits, vegetables, meat, dairy, and crops. This waste is produced both in the course of production and processing and it usually has some useful constituents, but the constituents in most cases, the cost of recovery, transportation and recovery of these substances is greater than their returns. The outturn of the waste however varies with the system of agriculture used and can be liquid, slurry or solid. Agro-waste is divided into the following: Animal waste: includes manure and dead animals. Food processing waste: for instance, in canning, only 20% of maize is put to use, leaving 80% as refuse. Crop residues: including corn leaves, sugar cane fibers, and peels of fruits and vegetables. Toxic waste: consists of pesticides, insect sprays, and weed killers. While there are no precise figures available, and such a waste is difficult to quantify, it is however estimated to be a significant proportion of the overall waste in developed societies. With the expansion of agricultural production, there has been an increase in the generation of waste consisting of animal waste, crop waste, and waste from agricultural industries. This tendency is also expected as agriculture practice becomes more advanced in the countries of the developing world. It is estimated that approximately 998 million tonnes of agricultural waste are produced every year as a rough average around the world. On farms, it has been assured that organic waste comprises approximately 80% of the overall solid waste generated, while that of livestock manure is estimated to range between 5.27 kg per 1,000 kg live animal weight each day.(Mamo and Zewide 2022)

The '3R' Approach to Agricultural Waste Management (AWM). One of the best methods used to address waste and its effects is the 3R approach. In other words it encompasses three key perspectives Reduce, Reuse, Recycle.

Reduce: this refers to making judicious use of the available resources so that there is little or no waste from the onset. Certain products and methods generate less waste and as such can help curb waste generation. Reuse: This entails using a material or even a component of a material several times before disposal. Waste products can be treated to render them reusable, thus saving on the need to extract for new resources. Recycle: A process in which by products are made into the same product or new ones. Waste resources are more of a raw material, created an extended existence of waste and addressed the problem of new materials.In applying the 3Rs, we conserve the earth's resources, lessen the waste generation and the negative effects caused by human activities. Items are not thrown away after use but rather are used for a different purpose, hence saving on new raw materials and energy. On the other hand, waste is processed and turned into useful items. Economic Benefits: The 3R approach can yield economic benefits in that waste can be recycled and turned into productions that earn revenue. Sustainability: Implementing the 3R system enhances sustainability in the aspect that it increases the utilization of available resources. Community Involvement: Getting people to buy into the 3Ring system brings people together to work towards better waste management.If the 3Rs are practised (individually or collectively), it promotes the responsible use of resources, enhances the usefulness of existing resources, and most importantly, avoids wastage with its attendant negative effects.(Koul *et al.*, 2022)

In the past in China, the communities adapted to the changing environment by turning the waste back into productive resources. Manure was collected from farm animals and applied to the farmland as organic fertilizer to improve soil quality. Wastes such as straw after harvesting were used for roofing covers, animal litter, and even cooked as firewood. These waste recycling practices practiced in the early times were out of need as most farmers strived to get the optimum yield from the scarce available resources.

The raw or semi-processed organic wastes which are not managed properly and not disposed of properly have a detrimental effect on marine and terrestrial environments. With the escalated adoption of mineral fertilizer by farmers, a meagre forty-seven percent of the arable land still receives any form of organic manure which used to supply a total of 18%, 28% and 75% of nitrogen, phosphorus and potassium respectively of the whole nutrient input in the year 2000. At present, on average, only 35% from all available nutrients (considering both inorganic and organic fertilizers) is provided to the cropland by organic manure sources in China.(Ju *et al.*, 2005)

Modern Innovations and Current Trends

In recent decades, there has been a resurgence in the utilization of agricultural waste, driven by environmental concerns and technological innovations. Advanced processes such as bioconversion, pyrolysis, and fermentation have opened new avenues for waste-to-energy projects and the production of bioplastics, biochar, and other eco-friendly products. Governments and organizations worldwide have implemented policies promoting the 3R approach—Reduce, Reuse, Recycle—to manage agricultural waste sustainably.

Agricultural solid waste (ASW) is a global issue but more concerns in the nations whose economy depends on agriculture such as India, China, Japan, Indonesia, Malaysia, e.t.c. There is so much agricultural waste such as crop residues, skins, leaves, corn cobs, rotten food e.t.c. then produced every year that is a threat to the environment and is not well disposed of. These experts assert that to effectively curb the menace of agricultural solid waste disposal, it becomes necessary to invent and most importantly put into practice techniques of waste, minimization, and recycling. The tends to be a concentration of these categories in waste materials. It is possible because of this structure to manufacture a variety of by-products, including feed for livestock, bio-insecticides, biofuels, organic fertilizers, single-cell products, nanoparticles, bioplastics, chitosan, antibodies, and the like. Besides, different strategies for developing agricultural solid waste into useful new products should also be encouraged. This review focuses on the origins of agricultural solid waste generation and guidance on the management of such wastes with the aid of modern practices for saving the environment and enhancing a country's economy. The result of our research is oriented towards the waste management approaching sustainability as it is unification of the best practices of waste management nowadays. This newly developed applies the novel practices developed in different sub-sectors of engineering to further environmental conservation. (Singh *et al.*, 2024)

Cultural practices affect how we live our daily lives including how we view food waste. Food loss or food waste has become a highly topical issue worldwide due to its consequences on the environment, health and economy. They cause environmental impacts and managing food that has been thrown away has cost implications in terms of energy and water. Therefore, wasting food emits greenhouse gases into the environment in the welfares of which concerns words like global warming. Food waste varies from culture to culture regardless of the country's development status.

Cultural variations present in cultural practices also bring about differences. These reasons have been put forward in the developed and developing countries concerning the reasons for excess food consumption. There is a wide range of considerations which may cause food waste, but at this point, it can be said without any doubt that in countries such as the U. K. and the U. S. who have upwardly mobile economies, food policies enhance wastage of crops in the farms significantly. In such developed nations, there is easy access to food and foodstuff is cheap too, therefore a certain way of perceiving food value and wasting food in the population is encouraged. Furthermore, the waste of food is being deliberately practiced in rich countries and that is alarming the global community. Elements such as age, consciousness and mindset are very essential in relation to the perception of food. One such example is the annual cultural fest 'La Tomatina' in

Spain, in celebration of which about 110–155 tons of tomatoes are wasted in what is billed as the biggest food battle in history. As a result other countries including Columbia, Costa Rica, China and Nevada follows suit and unfortunately it too comes with great environmental consequences.(Phasha *et al.*, 2020)

Agriculture Waste In Organic Fertilization Composting, Mulching, & Green Manure Benefits & Limitation

Organic Fertilization Composting :-

Agricultural activities, the ecosystem, and its denizens are interdependent. As a result, the quality of the environment is very essential for agricultural activities and there is a possibility that agricultural wastage management can affect human health and the surroundings negatively. In this regard, it is imperative for farmers to understand that they should not endanger the environment or people's health by treating, storing or the disposing of agricultural refuse in a careless manner. For instance, inappropriate management or disposal of organic waste can also pose danger to the environment and public health, for example, promoting disease proliferation and contaminating both land and underground water.

As per the European Union's regulations, the excreta of livestock is categorized as waste, which in this case, the farmer has to properly manage and dispose of by applying it to crops as a fertilizer without harming the environment. Pressure within As seen within the lower soils levels, due to un-vegetated surfaces and available water, any un-planted areas of the surface that have wide areas cover with wet fresh manure is likely to induce even temporary most non-cultivating soils aerobic processes like of denitrification or unintentional domseareach towards productive argillaceous captive sterile soils, ruinous soils in relevance to containment nutrients such bacteria, fungus, beneficials, weed seeds etc. Properly composted animal waste can be used as an organic amendment that may improve the physical and chemical properties of the soil.

Current research and industrial practice incorporates braiding of research fabrics as well as production with other agricultural bio-wastes, searching for alternative energies, and for the efficiency of the raw materials – the by-products and waste as well. One of the means of treating organic wastes is through the recycling of these wastes into organic products by means of pelleting them. Granular manure is a generalized name for all forms of organic fertilizers with all macro and micro trace elements combined into one. Pelleted manure's nutritive value is perfect for soil micro flora as it has a short solubility in water and is readily taken up by plants. Granular manure can be referred to as a grade of fertilizer because the integrated process of water removal reduces the size of at least 10 times the overall volume of the material packages as waste. (R. Mielda ys *et al.*, 2016)

Cover crops are essential for the success of organic agriculture, and in well-Designed rotation systems, they are capable of providing crucial ecological services that support agroecosystem sustainability. Without a doubt, the practice of planting soil is an efficient method of controlling the growth of weeds as well as eroding soils and leaching nutrients, while at the same time enhancing organic matter of the soil, maintaining permanent soil fertility and productivity.

Organic fertilizers and amendments are prepared from materials that are of animal and plant origin and properly processed as per the applicable organic certification guidelines. Two processes out of the approved transformation processes of organic materials to organic fertilizers are aerobic and anaerobic digestion. The biological treatment of organic material, using oxygen, results in a stable and organic enriched humic like end product that characteristically tends to mineralise at a very low rate in the soils. The material produced by aerobic fermentation of the digested mass or bacillus ulmus is a horticulture us form of nitrogen compound which is in chemical clouds at low weights and looking at their source materials can provide N and other elements within composting rates. Several studies have shown promising yield results from the application of compost, municipal solid waste and anaerobic digestates) for different crop species.(Montemurro *et al.*, 2013)

The composting process contributes to the preservation of underground water from contamination when compared to the landfilling waste management method, which has the potential to contaminate the freshwater aquifers. This is as a result of the cut down in the number of disease-causing organisms and harmful chemicals in the waning process. These are the disease-causing organisms found in waste that are dangerous for human health. The chemical-absorbing useful microbes contain such chemicals as persistent organic pollutants and endocrine disrupting chemicals during the process of composting, that were present in the soil before. Further, details on this are presented in this article. Use of composting, increases agricultural productivity and organic matter content of the soil due to the fact that composted materials are nutritious and their active ingredients include plant growth promoting organisms.

This goes a long way in ensuring food security. Aside from its use as a fertilizer, compost has a potential in the control of plant diseases and pests, erosion control and prevention, pollution control, weed control, landscaping, as well restoration of wetlands. It is also proactive since it promotes soil diversity and less dependence on chemical fertilizers which pose threats to the environment. Composting is a process that, from the very beginning, has been practiced under specific conditions, unlike a process that is natural and uncontrolled as it happens in the open environment. This is the process that deprives the definition of the activity of composting into decomposition which is a biological process without any control input. While composting is very useful, it takes quite a time for its readiness, also it may smell bad, even longer mineralization period, some pathogenic organisms are present within it that can resist temperature up to a level i.e. thermotolerant organisms and low level of nutrients provided.(Ayilara *et al.*,2020)

Mulching

Usage of mulch is beneficial not only in controlling the existing weed but also in preventing any weed from sprouting. A layer of mulch blocks sunlight. Young weeds are unable to photosynthesize and produce sugars necessary for growth without higher sunlight and hence perish. Some weed seeds need light in order to germinate which also mean that mulch can prevent germination. Lastly, generally small seeded weeds are unable to break through a 3 inch layer of mulch.

Due to the growing population and extensive urbanization, there has been an increase in demand for agricultural products but with land being a limiting factor. In many developing nations, the "primitive" technique of shifting cultivation has been replaced by the continuous practice of growing crops on the same piece of land, thus leading to a steady decline in the soil fertility and crop production. The harsh climatic conditions have also worsened the problem of soil fertility in the developing countries. Although organic fertilizers provide essential nutrients in the soil, their habitual application leads to depletion of soil organic matter, soil hardening and lowering of soil physical quality in general over time. Jean-Michel Lagrange Professor Organic farming can be said to rely heavily on soil organic matter (O.M) accumulation and biological activity as the two major sources of nutrients in soil which however depends largely on the addition of plant tissue. The decay of organic matter enriches the soil with nutrients for then following crops. Organic matter decomposition is governed by the size and structure of soil microbial biomass. It has been observed that mulching reduces evapo-transpiration process significantly and therefore it is advised for all the home garden crops including but not limited to potatoes, sweet potatoes, carrot, fluted pumpkin, pumpkin tomatoes and ginger.

Benefits of Mulch has many positive effects on soils. Generally, organic mulches conserve water, reduce weeds, improve soil quality and enhance plant growth.(Ibeawuchi *et al.*, 2015)

Appropriate addition of materials is often employed before, during or shortly after sowing operations, which is also applied and spread over the soil surface such as plastic, pour soil, crop remains, animal dung, sands, stones, cement among other things. The major reasons why mulching is done includes limiting moisture/ water loss or erosion, elevating or regulating soil temperatures, modifying soil moisture storage, and controlling weeds. This has helped to improve crop production, promote growth of plants and conserve water.

Types Of Mulching :-

Mulching materials can be classified into three major groups; organic, inorganic and special materials. Organic mulching materials are made from Agricultural wastes, wood industrial wastes, processed leftovers and animal manures. Inorganics materials use things such as polyethylene plastic films and synthetic polymers. There are also several new types of eco friendly materials that have come out such as new biodegradable or photodegradable plastic films, surface coating and flexible biodegradable polymer films so as to ease the way of doing things.

Straw

Post-harvest, straw is or crop leftovers are abundantly available. Straw mulch is a very light and easy to apply and use material. Paddy straw today is widely used for field mulching as it facilitates better crop growing conditions. In contrast, when it is applied as mulch, straw may create several challenges. Straw mulches have to be changed on an annual basis. Because they are extremely flammable and include grain seeds that could germinate and deplete soil nitrogen levels as they decompose.

Bark Mulches

Such mulches are efficacious as they have a greater moisture retention capacity over a prolonged period and they also stretch the time when the moisture is available for crop use. Often, wildflower sods or mats are placed over the soil for such purposes. Nevertheless, it has high acidity and should be avoided in growing vegetable gardens. While this mulch should not be employed on beds it is great for walkways in between the beds instead.

Wood Chips

Wood chips are produced from waste wood and different types of trees. However, wood chip mulches have a high C:N ratio meaning that nitrogen in the soil may not be easily available for uptake by plants while they break down.

Sawdust

In places where sawdust is easily accessible, it is commonly made the more. It is generated from various wood finishing activities. Its nutritional content is less compared to straw since it contains only fifty percent of the nutrients. The decomposition process is very gradual due to the high C:N ratio. This will cause nitrogen 2 deficiency in the soil during its decomposition, making it essential to apply fertilizers frequently. Due to its sour nature, it should not be applied on any soil with a low ph level. It does, on the other hand, hold water for a relatively long time.

Compost

Mulch and soil enrichers at their best, compost can be fashioned at home with relative ease, using waste products ranging from leaves, straw and grass to plant ends. The use and importance of compost in agriculture is already an age-old practice. It enhances the structure of the course, more so the organic matter content which in turn increases the water holding capacity of the soil, thus promoting soil health. Owing to the relatively high nitrogen content of the compost, it is rather ineffective in application on plots growing vegetables as it encourages excessive weed growth.(Beltagi *et al.*, 2022)

Green Manure

Agricultural approaches have an immediate effect on the ecological balance and alter the basic elements that support the environment. After several years of cultivation, altered soils reveal transformations in the physical, chemical, and biological attributes, where the soil's organic components perform a critical function in assessing the restoration imbalance because this is associated with encompassing the cation exchange capacity, aggregate sturdiness, and nutrient distribution, in an adverse manner. In the quest for sustaining agriculture, the term green manuring comes into picture which is a practice of growing plants either in rotation with or intercrops with crops of economic importance leading to improvement in the properties of the cultivated soil and agriculture production indirectly enhanced. Green manures have numerous advantages. These include: increased organic matter in the soil, reduced evaporation and erosion caused by crop debris, nutrient cycling and breaking up of hard pans. Other important gains are increased soil microorganisms, lowered vectors and pathogens and reduced weed infestations.(Zaccheo *et al.*,2016)

Green manure crops are those which can be nutrition providing crops for fully grown economic crops and help in enhancing the productivity of soils; this mostly includes leguminous crops which have rhizobia heterotrophs that facilitate nitrogen fixation as seen in crops like Chinese milk vetch and alfalfa, and crops that do not belong to the legumes family for example rye grass and barley. It has been known since before the advent of modern agriculture that in countries such as China, India and Japan, ice cultured leguminous green manures were grown for wetland rice cultivation as an important source of N. There are some incomplete representations of information saying that in the period from the late 1950s till the early 1980s the number of cultivated land areas under green manuring practices reduced, this could be due to the changes in social systems, economic structural advances and modes of agricultural practice especially in China. Non organic fertilizers are now the most used nutrient input into the soil for crop production. Nevertheless, this advantage has become evident in a very short period of time owing to the negative soil fertility impacts, farm non point source pollution, and issues of soil acidity arising from high usage rates of chemical fertilizers over decades, the implementation of environmental agriculture service projects, and due to the new awareness of green manures fostered among people in recent years.(Lei et al.,2022)

Agricultural Waste in Animal Feed Use of crop residues and processing by-products in livestock diets Nutritional considerations and safety issues:

World food production is undergoing strong intensification, generating large amounts of food co-products and wastes, which partly drives current social and environmental pressure for the effective reuse of agricultural industry residues (Cooper *et al.*, 2013). The use of agro-industrial by-/co-products in farm animal nutrition minimizes food industry environmental footprint and increases profitability and valorization of the agricultural by-products as feeding food residue to livestock is a very efficient way of upgrading low quality materials into high quality foods (Waldron 2007).

In addition, the production of these unwanted materials results in growing disposal and possibly significant pollution challenges as well and a wastage of biomass and nutrients. Industrial ecology and circular economy are also described as the prominent paradigms for eco-innovation that aim at a zero waste society and economy (Mirabella 2014).

In addition, based on the rapid increase of per capita incomes and urbanization in developing countries, the global feed ingredients market is likely to expand at the compound annual rate of 3.8% during these years and predicted to attain near $20 million by 2018. Finally, the phytonutrients production via biorefinery and new extraction processes has met an increasing availability of phytochemicals and other interesting target compounds that can be added to several products as functional additives (Galanakis *et al.*, 2012). Agri-food by-/co-products from fruit and vegetable processing are attractive raw shelters of phytochemicals (carotenoids, phenolics, and flavonoids), antioxidants, antimicrobials or vitamins and dietary fat displaying desirable technological functions or declared nutritional properties. In some other key review studies the applications including fruits and vegetables residues or their beneficial extractable compounds in food industry were presented (Laufenberg *et al.*, 2003). The most widely, and extensively studied, feed ingredients in animal nutrition from the fruit and vegetable processing residues have shown that they can influence animals performance. On the other hand, by-/co-products from fruit and vegetable processing as sources of phytochemicals represent a more recent category of feed ingredients and little is known concerning their potential applications, functionalities as well as biological activity, bioavailability (Ayala-Zavala *et al.*, 2011).

Feed Additives for Improved Product Quality Functional Ingredients from Fruit and Vegetable Co-Products:

In accordance to the European Food Safety Authority (EU 2008), feed additives for improved animal product quality can "favorably affect":

a. the sensory characteristics and acceptance of the products, i.e., antioxidants and colorants.

b. the nutritional value of the products, i.e., long chain polyunsaturated fatty acids, conjugated linoleic acid (CLA).

c. the microbial quality of the products.

For example, tomato skin contains higher levels of lycopene compared to tomato seeds (Knoblich *et al.*, 2005). In addition, the sodium content of tomato skins is increased by using sodium hydroxide to mechanically peel the tomatoes.

Furthermore, the names of by-products (residues, pulp, cake, waste, etc.) may not accurately describe the composition and composition of the product. Therefore, the chemical composition of by-products is determined before application in animal feed.

Food Feed Food System Description:

The current food system is shown in Figure 1. In the case of junior production (such as olive leaves or pomas), a common product is collected in the fruit and vegetable industry. They are then used as unprocessed surpluses or treated. Processing procedures may include drying as most of these materials have high moisture content leading to product spoilage, or may require advanced processing/biorefining techniques to capture specific compounds such as phenols, vitamins, fatty acids or carotenoids. Citrus fruit has a moisture content between 76% and 83%, olive cake has a moisture content of around 30%–50% and grape pomace has an average moisture content of around 64% (Makris *et al.*, 2007).

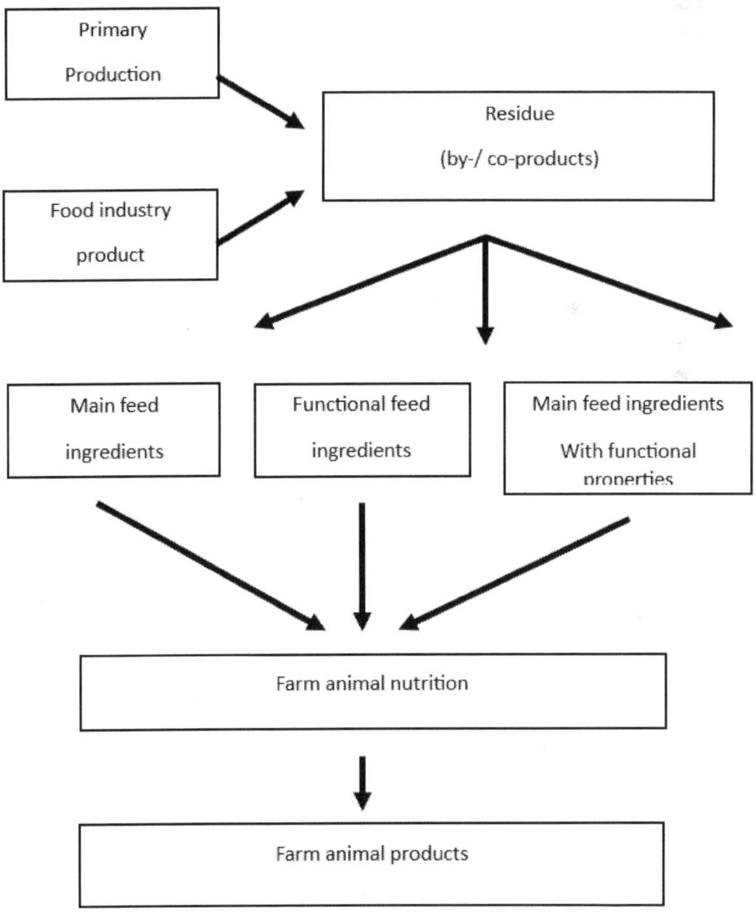

Fig No. 1: Simplified schematic diagram of food feed food system.

Quantitative and qualitative value of agro-byproducts as animal feed:

Many agro-industrial by-products can be fed directly to animals without any modification or used after fermentation of agricultural residues. Generally, by-products that can be used as feed must be economical, nutrient-rich, and free of toxins or other substances that may be unhealthy for animals. The feed situation in Europe has shown that to maintain the animal population at the current level, the protein deficit is in the range of 25-28 million tons and varies from several hundred thousand to millions of tons in different countries (Szebiotko, 1984). Denmark imports 30% of the total amount of protein used to cover the country's protein needs, even with developed agriculture and high levels of animal husbandry. Agricultural by-products refer to by-products obtained from industry by processing major products in large quantities. They are less fibrous, more concentrated, have higher nutritional value and are cheaper than crop residues (Aguilera, 1989).

1. Plant source
2. fermentation industry.

Current limitation for utilization of agro- byproducts as animal feed:

There are many factors such as protein concentration and its biological value, digestion, energy levels, fatty and carbohydrates, amino acids, vitamins and mineral content. Non-traditional feed resources, such as pesticide residues and by-products, are not fully or underutilized for various reasons. The production of pesticide residues is diffuse and seasonal and in some cases of lower quality, especially processing grades. The cost of agricultural products is uncompetitive and cheap for farmers (Ravindran & Blair 1992).

Traditional Uses of Agricultural Waste in Small-Scale Crafts and Fuel

Summary of how traditional agricultural waste is utilized in small-scale industries and crafts. Significance of eco-friendly methods and advantages for the local economy. Short overview of traditional crafts and fuel usage, providing background for in-depth conversations. Recently, there has been a growing interest in the utilization of agricultural waste in traditional crafts and small-scale industries for its potential role in sustainable development (Deshmukh, Rajput, Das, & Alam, 2024).

Historical Background Of Agricultural Waste In Traditional Crafts:

Dhokra art, an ancient form of metal casting in India, exemplifies the historical use of agricultural and natural materials, dating back 4,000 years and incorporating local resources like beeswax and bell metal . Examination of historical contexts, such as Dhokra art and other metal crafts, which use natural materials and agricultural residues. (Deshmukh *et al.*, 2024).

Role of Agricultural waste in preserving Cultural Heritage and Traditional Crafts:

Utilizing agricultural waste is crucial for generating income in rural areas, as it sustains livelihoods in areas where traditional handicrafts are the main source of employment (Deshmukh *et al.*, 2024).

Agricultural waste as a source of Solid Biofuels:

Studies have demonstrated the significant potential of solid biofuels, particularly in SMEs, to replace fossil fuels in industries like brick and lime kilns. Agricultural waste like rice husks, wheat straw, corn stalks, and sugarcane bagasse has long been utilized worldwide for crafting various items. The versatility, durability, and environmentally friendly properties of the materials are highly appreciated. In numerous cases, these materials were reused to create useful or decorative items, frequently employing traditional and basic methods. (Ruíz-Carmona and colleagues, 2021).

Basketry and Weaving: A common utilization of agricultural waste is in the crafts of basketry and weaving, using materials such as rice husks, palm fronds, and bamboo. In India, rural artisans have been making baskets, mats, and ropes for a long time using agricultural byproducts like rice straw or wheat straw. These materials are light, pliable, and durable, making them perfect for such projects (Gupta, 2011).

Paper and Pulp Crafts: Traditional paper-making methods have also made use of agricultural waste like straw, banana fibers, and sugarcane bagasse. This procedure includes breaking the fibers apart and converting them into pulp that can be dried into sheets for writing, printing, or crafting. A study by Inoue et al. (2016) emphasizes the use of agricultural waste, such as bagasse, as a cost-effective and eco-friendly substitute for wood pulp in making paper.

Textile Production: Certain societies utilize leftover agricultural materials in the creation of textiles. An example is the use of jute, obtained from the jute plant, a crop usually seen as a by-product after fibre extraction, in Southeast Asia ,to produce ropes, sacks, and rugs (Kabeer, 2017). This method is still commonly used today, with contemporary variations that involve utilizing banana and hemp fibres.

Materials and Techniques in Traditional Handicrafts:

The utilization of agricultural waste in traditional crafts and as a fuel source has been a common practice in different societies. These methods not just promote a sustainable way of utilizing waste byproducts but also protect cultural heritage and offer useful solutions to daily needs. This section examines the methods and materials used to incorporate agricultural waste into small-scale crafts and fuel production, drawing on scholarly research to validate these claims.

Applications of Biofuels in Small-Scale Industries:

In Mexico, small and medium enterprises have started using solid biofuels like bagasse and wood chips from agricultural residues for energy-intensive processes, decreasing their dependence on fossil fuels (Ruíz-Carmona *et al.*, 2021)

Case Studies of Small-Scale Handicrafts using Agricultural Waste:

Detailed examination of local artisanal traditions, such as Indian bell-metal crafts and the incorporation of agricultural byproducts in crafts from Southeast Asia. Evaluation of Dhokra artisans' SWOT analysis focuses on strengths, weaknesses, opportunities, and threats concerning the utilization of agricultural waste.

Economic Viability Of Traditional Crafts Using Agricultural Waste:

Examination of revenue and job creation in countryside areas. Economic benefits generated from agricultural waste, particularly in areas with scarce resources.

Agricultural Waste As A Source Of Solid Biofuels:

Summary of how agricultural waste is transformed into solid biofuels like firewood, pellets, and briquettes. Importance in replacing traditional energy sources for small businesses, specifically in rural areas.

Agricultural Waste As Fuel:

Agricultural waste is also utilized as fuel in numerous traditional societies. This functions as both a method of getting rid of waste and a useful answer for cooking and heating requirements. Common types of agricultural waste utilized as fuel are rice husks, corn stalks, and wood chips, which can be burned as is or turned into biofuels.

- **Biomass Fuel:** Agricultural waste and other biomass materials are burned in traditional stoves, offering a cheap and readily available energy option. Research conducted by Gupta *et al.* (2014) suggests that utilizing rice husks as biomass fuel in rural India decreases waste and offers a sustainable cooking energy option for villageswithout access to modern fuels such as LPG or electricity.
- **Briquetting** of agricultural waste is a modern method that is becoming more popular. During this procedure, rice husks, sawdust, or sugarcane bagasse are compacted into solid blocks for use as a cleaner and more effective fuel option. A study conducted by Sharma *et al.* (2019) discovered that briquettes produced from agricultural waste release lower levels of pollutants in comparison to traditional fuels, presenting a more eco-friendly choice for rural families.
- **Making biochar:** Biochar is created by heating agricultural waste through pyrolysis, turning it into a type of charcoal used as fuel. The procedure

includes heating the biomass in a low-oxygen setting, transforming it into a durable, carbon-heavy material that has applications as both a fuel and a soil amendment. In a study conducted by Demirbas (2011), it was found that producing biochar from agricultural waste not only resolves waste disposal problems but also aids in reducing greenhouse gas emissions.

Techniques in Crafting with Agricultural Waste:

The methods utilized for transforming agricultural waste into practical items usually require basic tools and simple techniques, depending on manual skills and traditional wisdom handed down over the years. Common methods used in creating products from agricultural waste.

- **Drying and stripping:** Numerous agricultural byproducts undergo drying to lower moisture levels, enabling easier handling and increased durability. An instance of this is when wheat straw is frequently dried in the sun before being converted into slender fibers for weaving or binding purposes (Kothari, 2015).
- **Forming and Modifying:** Mixing agricultural waste like clay and cow dung with rice husks or straw has been utilized to produce mud bricks or plaster for construction in rural regions of Africa and Asia. These shaped mixtures are subsequently air-dried or fired in furnaces to create long-lasting materials for buildings (Mingliang *et al.*, 2020).
- **Binding and Coiling:** In basket making, coiling is a common technique that involves wrapping long, pliable pieces of straw or reeds into spiral forms and securing with threads made from the same type of plant material. This method enables the development of robust, long-lasting baskets, mats, and other woven items (Basu *et al.*, 2013).

Environmental and Economic Impacts:

The use of agricultural waste in small-scale crafts and fuel has several environmental and economic benefits:

- **Sustainability:** Utilizing agricultural waste for crafting and fuel production decreases the demand for new resources like timber and reduces the environmental impact of waste disposal (Gosh et al., 2018).
- **Economic Opportunities:** In numerous rural areas, using agricultural waste in crafts can offer additional income opportunities for families. Artisans can produce sellable items by working on small, low-tech projects and making use of materials that would otherwise be thrown away.
- **Waste Reduction:** Using agricultural waste for crafts and fuel reduces waste volume, easing pressure on landfills and promoting cleaner environments.

Environmental Benefits and Challenges of Agricultural Waste Utilization:

Impact on the environment when reusing agricultural residues for crafts and fuel purposes. Tasks such as resource depletion, emissions, and waste management. Sustainability models for handling environmental consequences.

Government Policies and Support for Traditional Crafts and Biofuel Use:

Summary of governmental and non-governmental efforts backing traditional crafts and biofuel usage. Policies encouraging environmentally-friendly practices, like "Make in India" and rewards for biofuel utilization.

Sustainability Framework and Future Directions:

Incorporating sustainability into the utilization of agricultural waste in small-scale operations, with a primary emphasis on environmental, social, and economic aspects. Prospective pathways for traditional crafts and biofuels, involving technological progress and policy suggestions.

Conclusion

Agricultural and food industry waste accounts for a large proportion of global agricultural productivity (approximately more than 30%). These wastes mainly include fruit and vegetable waste, sugar industry waste, etc. They are valuable biomass and potential solutions to the problems of animal nutrition and global protein and calorie supply if appropriate technologies can be used for their valorization through protein enrichment.

Technologies developed for better utilization should take into account factors such as the characteristics of individual waste and the environment in which it is generated, processed and used. The use of such technologies especially requires that products be delivered that are not only safe for feed use, but also safe from a human feeding perspective.

The development of the agricultural products processing industry should be based on fundamental and applied research in food science and processing technology. Local studies are needed to understand processing facilities, processing conditions, chemical and nutritional diets and their effects on animal performance.

The use of nutrient-rich agricultural products in animal feed production will reduce food costs and animal product prices. It will also help achieve nutrient self-sufficiency through local waste. Systems for the production of feed from waste provide a vertical, mutually complementary integration of animal husbandry and in turn solve the problem of waste and environmental pollution.

The use of agricultural waste in animal feed will create a solid foundation for continued technological development and management of the livestock industry.

By-products from fruit and vegetable processing remain an underutilized source for supplementing livestock with functional compounds and producing value-added products. The commercial use of by-products of the fruit and vegetable industry as functional feed ingredients presents challenges and opportunities for field researchers.

CHAPTER 4

TECHNOLOGICAL INNOVATION IN WASTE VALORIZATION

Widescale processes which could generate valuable resources from waste, have become crucial in decreasing pollution and ensuring sustainability. Based on the current technological changes, this chapter focuses on the key biochemical, thermochemical, mechanical and physical, and advanced waste treatments. It presents the concept, applications, and innovations of each group. Some of the recent developments about the future perspective of technologies, overcoming challenges in waste valorization technology, are outlined at the conclusion.

Background and Scope Introduction

The growing population coupled with rapid industrialization are adding enormous amounts of waste leading to a significant threat posed by the waste to environment. Landfills are rising day by day, whereas the pollutants from the dumpsites are polluting atmosphere, water, and even the soil. To conquer this challenge, waste valorization has emerged as the paramount one. Waste valorization is the process of transforming waste materials into valuable resources, including energy, fuels, chemicals, and materials, thus minimizing environmental impact and promoting a circular economy. Technological innovations have greatly increased the scope of waste valorization, offering innovative solutions for various waste streams. This chapter presents a comprehensive review of key technological innovations in waste valorization, categorized into the following:

Biochemical Processes: Composting, anaerobic digestion, fermentation, and bioenergy production.

Thermochemical Processes: Pyrolysis, gasification, incineration, and biochar production.

Mechanical and Physical Processes: Milling, pressing, extraction techniques, and the creation of fiber-based products and bio-materials.

Advanced Waste Treatment Technologies: Biotechnology, nanotechnology applications, and smart waste management systems.

Biochemical Processes

Biochemical processes utilize biological agents, mainly microorganisms, to convert organic waste into valuable products. These processes are often characterized by mild operating conditions and the potential to produce a wide range of bio-based products.

Composting

Composting is the natural process of aerobic decomposition of organic materials, such as food waste, yard trimmings, and agricultural residues, into a nutrient-rich soil amendment called compost (USEPA, 2023). This process is carried out in a controlled environment to maximize microbial activity, thereby stabilizing and sanitizing the organic material (Zhang et al., 2018).

Process Overview

- The composting process generally involves the following steps:
- Preparation: Organic waste is shredded or chopped to increase surface area and facilitate microbial activity.
- Mesophilic Phase: Microorganisms thrive at moderate temperatures (30-40 C), breaking down easily degradable organic matter.
- Thermophilic Phase: As microbial activity increases, the temperature rises to 55-65 C, killing pathogens and accelerating the decomposition process.
- Curing and Maturation: The compost is left to cool and mature, during which time microbial activity slows down and the compost stabilizes.

Applications

Compost is widely used as a soil amendment to improve soil structure, enhance water retention, and increase nutrient availability for plant growth. It can also be used for landscaping, erosion control, and greenhouse cultivation.

Advantages

- Reduces the volume of organic waste
- Produces a valuable soil amendment
- Reduces greenhouse gas emissions from landfills
- Promotes sustainable agriculture

Challenges

- Has to be managed carefully about moisture content, aeration, and temperature.
- May produce unpleasant odor if not managed properly.
- May require extensive land area for composting facilities.

Anaerobic Digestion

Anaerobic digestion is the breakdown of organic matter with microorganisms in the absence of oxygen, resulting in the formation of biogas that is primarily methane and carbon dioxide (Appels et al., 2011). Biogas is used as renewable energy in the generation of electricity or heat or fueling of vehicles (Sari et al., 2014).

General Description

The general four steps in anaerobic digestion process include:

- **Hydrolysis:** The complex organics are degraded into small molecules, including sugars and amino acids.
- **Acidogenesis:** Acid-tolerant bacteria convert simpler sugars into organic acids in the form of acetic acid and propionic acid.
- **Acetogenesis:** Acetogenic bacteria convert organic acids to acetic acid, hydrogen, and carbon dioxide.
- **Methanogenesis:** Acetic acid, hydrogen, and carbon dioxide are converted by the methanogenic bacteria to methane and carbon dioxide.

Applications

Anaerobic digestion may be able to be adapted to utilize any organic waste flow, that includes food wastes, wastes from animals manure to agricultural residues, and utilized in wastewater and sludge stabilization.

Benefit

- Produces renewable energy
- Reduces greenhouse emission in landfills
- Decreases organic waste volumetric and toxicity
- Can facilitate wastewater treatment

Challenges

It needs specific conditions to be kept anaerobic. It can be prevented by the existence of specific contaminants, including heavy metals. It needs a substantial capital investment to build up large-scale facilities.2.3 Fermentation

Fermentation is the metabolic process through which microbes convert carbohydrates into organic acids, alcohols, or other chemicals (Soccol et al., 2010). The main application of fermentation involves several bio-based products in industries, such as production in bioethanol, lactic acid, and further value-added chemicals derived from organic waste streams (Mussatto et al., 2011).

Process Overview

A general overview of a common fermentation process typically consists of the following steps:

- Substrate Preparation: Organics waste undergo pretreatment to make them susceptible for microbial conversion.
- Inoculation: The substrate is inoculated with microorganisms, such as bacteria or fungi.
- Fermentation: The microorganisms metabolize the substrate, producing the desired products.
- Product Recovery: The desired products are separated from the fermentation broth.

Applications

Fermentation can be used to produce a wide range of bio-based products from organic waste, including:

- Bioethanol: Produced from lignocellulosic biomass, such as agricultural residues and wood waste.
- Lactic Acid: Produced from diverse carbohydrate sources such as whey, corn starch, and sugar beet molasses.
- Other Chemicals: Produced from organic waste streams including butanol, acetone, and organic acids.

Benefits

- Makes use of renewable resources
- Produces useful chemicals and materials
- Minimizes the use of fossil fuels
- It can be coupled with other technologies that valorize wastes

Challenges

- Can be susceptible to variations in environmental conditions.
- Can sometimes necessitate process parameters to be optimized in order to have a good yield.
- Susceptible to contamination by unwanted microorganisms.

Bioenergy Production

Bioenergy production refers to the conversion of organic waste into different forms of energy, such as biogas, bioethanol, and biodiesel. These biofuels can be used as substitutes for fossil fuels, which reduces greenhouse gas emissions and promotes energy independence.

Biogas Production

Biogas is produced from the anaerobic digestion of organic waste, which has been discussed above. It can be used for several applications:

- Electricity generation
- Heat production
- Vehicle fuel

Bioethanol Production

Bioethanol is produced through the fermentation of sugars derived from various sources, including:

- Sugarcane
- Corn

Lignocellulosic biomass

Bioethanol can be blended with gasoline to improve fuel efficiency and reduce emissions.

Biodiesel Production

Biodiesel is produced through the transesterification of vegetable oils or animal fats. It can be used as a direct replacement for diesel fuel in diesel engines.

Advantages

- Reduces dependence on fossil fuels
- Reduces greenhouse gas emissions Improves energy security
- Enhances rural development

Challenges

- Can compete with food production for land and resources.
- May require huge investments in infrastructure.
- Exposes itself to feedstock supply volatility and price fluctuation

Thermochemical Processes

Thermochemical processes convert waste to energy or valuable product using thermal treatment. Such process can be applied to virtually all types of waste, such as municipal solid wastes, industrial wastes, agricultural residues, etc.

Pyrolysis

Pyrolysis is a thermochemical process, in which organic materials are thermally decomposed in the absence of oxygen, to yield bio-oil, biochar, and syngas (Mohan et al., 2006). Bio-oil can be further processed into transportation fuels or chemicals, whereas biochar can be applied as a soil amendment or adsorbent (Lehmann et al., 2011).

Process Overview

Pyrolysis typically involves heating organic waste to high temperatures (400-800 C) in the absence of oxygen. The process can be carried out in various reactors, including: Fixed-bed reactors Fluidized-bed reactors Rotary kilns 3.1.2 Products Bio-oil: A liquid product that can be used as a fuel or a feedstock for chemical production. Biochar: A solid carbon-rich product that can be used as a soil amendment or adsorbent.

Syngas: It can be utilized to generate electricity or chemical production.

Sources and related content

Waste Valorization: Transforming Challenges into Opportunities

Applications

Pyrolysis has been applied to a great variety of waste streams, for example:

- Municipal solid waste
- Industrial wasteAgricultural residues
- Tires
- Plastics

Benefits

- Conversion of a wide range of waste streams into valuable products is possible.
- Relatively clean fuels and chemicals.
- Biochar, a valuable soil amendment, can be produced.
- Decreases the volume and toxicity of wastes.

Challenges

- Requires large temperatures and specialized equipment.
- Complex mixtures of products that require further processing can be produced.
- May need considerable energy input to facilitate the heating process.

Gasification

Gasification is a thermochemical process where organic material is converted into a combustible gas mixture, syngas, through partial oxidation at high temperatures (Brandt et al., 2009). Syngas can be used to produce electricity, heat, or as a feedstock for chemical synthesis (Kurkijärvi et al., 2007).

Process Overview

The following steps are typically involved in the gasification process:

- **Drying:** Organic waste is dried to reduce moisture content.
- **Pyrolysis:** The dried waste is heated in the absence of oxygen, which produces volatile gases.
- **Oxidation:** The volatile gases are partially oxidized with air or oxygen to produce syngas.

Products

Syngas: A mixture of carbon monoxide, hydrogen, carbon dioxide, and methane.

Applications

Syngas can be used for:

- Electricity generation
- Heat production
- Chemical synthesis (e.g., methanol, ammonia)
- Fuel cells

Advantages

- Can be used to convert a wide range of waste streams into a versatile energy source.
- Produces a relatively clean fuel.
- Can be integrated with other energy systems.

Challenges

- Requires high temperatures and specialized equipment.
- Can produce tar and other contaminants that require removal.
- May require significant energy input for the gasification process.

Incineration

Incineration is a high-temperature thermal process that involves the combustion of waste materials to reduce their volume and generate heat or electricity (USEPA, 2021). Although incineration may be effective in reducing the volume of waste, air pollution control measures must be implemented to minimize the release of harmful pollutants (Wärnhammar et al., 2003).

Process Overview

Incineration is usually burning the waste materials at high temperatures (800-1100 C) in a controlled atmosphere. The heat produced from combustion can be utilized in the generation of steam that can be used for generating electricity or other industrial purposes.

Benefits

- Reduces the volume of waste greatly.
- Can produce electricity or heat.
- Can be used to extract metals from waste.

Problems

- It has the potential to produce air pollution, including dioxins, furans, and heavy metals.
- It requires strict air pollution control measures.
- Public perception concerns over possible environmental and health effects.

Biochar Production

Biochar is a carbon-rich solid byproduct of biomass pyrolysis. It can be applied as a soil amendment to enhance soil fertility, improve water retention, and sequester carbon (Lehmann et al., 2011).

Overview of the Process

Biochar is usually produced through slow pyrolysis of biomass at temperatures ranging from 300-700 C in the absence of oxygen.

Applications

Biochar can be applied for:

- Soil amendment to increase fertility, water retention in soil
- Carbon sequestration
- Air and water cleaning
- Animals feed

Benefits

- Improves the quality of soil and agricultural productivity increases
- Sequesters carbon and reduces climate change impact
- Can be applied to many environmental purposes

Problem

- Needs proper management not to pose negative impacts on the environment.
- Further research may be necessary to properly optimize biochar production techniques and application.

- Mechanical and physical processes are used to separate, sort, and transform waste materials intovaluable products through physical means. Such processes are often used as preliminary steps in waste valorization chains to recover valuable materials and prepare waste streams for further processing.

Milling, Pressing, and Extraction Techniques

Milling, pressing, and extraction techniques are used to separate and recover valuable materials from waste streams.

Milling: It reduces the size of solid waste to make further processing easier (Kumar et al., 2013).

Pressing: Compacts materials for volume reduction and liquid removal (Kamm et al., 2006).

Extraction Techniques: Includes solvent extraction and supercritical fluid extraction, which separate certain constituents from waste material (Chemat et al., 2012).

Production of Fiber-Based Products and Bio-materials

Fiber-based products, like paper and cardboard, can be recycled and reused several times (European Commission, 2018). Furthermore, new bio-materials, including bioplastics and biocomposites, are being designed from renewable resources, thereby providing sustainable alternatives to conventional petroleum-based materials (Auras et al., 2011).

Advanced Waste Treatment Technologies

Advanced waste treatment technologies use the most current scientific principles to enhance processes of waste valorization. These technologies provide promising routes for enhancing the efficiency of waste treatment and the discovery of new valorization routes.

Applications of Biotechnology and Nanotechnology

Biotechnology

This is the application of microorganisms or enzymes for the degradation or transformation of wastes (Banerjee et al., 2017).

Nanotechnology: Facilitates the design of advanced materials and processes for waste treatment such as nanofiltration membranes and nano-catalysts (Kumar et al., 2018).

Intelligent Waste Management Systems

Smart waste management systems employ information and communication technologies such as sensors, GPS, and data analytics to optimize collection, transportation, and processing of wastes (Goyal et al., 2017). These systems permit real-time monitoring of streams of waste, predictive equipment maintenance, and efficient allocation of resources, thereby making better waste management outcomes achievable (Caputo et al., 2019).

Challenges and Future Perspectives

Despite the fact that great strides have been made in waste valorization technologies, several challenges persist:

- Economic viability: Most of the technologies for waste valorization demand heavy capital investment and might not be economically viable in some contexts.
- Technological inadequacies: Certain technologies may have limitations with regards to efficiency, scalability, and applicability for a particular waste stream.
- Public perception: Public fears and apprehensions about potential adverse environmental and health effects associated with a particular waste valorization technology are likely to act as barriers.
- Integration and coordination: Effective integration and coordination among various waste valorization technologies and stakeholders are very important for successful implementation.
- Future research and development activities should be focused on the following:
- Development of cost-effective and efficient waste valorization technologies.
- Improvement in the sustainability and environmental performance of existing technologies.
- Addressing public concerns and public acceptance of waste valorization initiatives.
- Promotion of collaboration between researchers, industry, and policymakers to hasten the development and deployment of innovative waste valorization solutions.

Conclusion

Technological innovations are increasingly playing a pivotal role in advancing waste valorization practices where waste, once a liability, becomes a valuable resource.

Biochemical, thermochemical, mechanical and physical, and advanced waste treatment technologies will be integrated together to make a more sustainable and circular economy with lesser waste and more resources being maximized. Continued research and developments in these areas are called for to address the increasing burdens of waste management and forge a more sustainable future forward. Waste valorization--the conversion of waste to valuable resources--has therefore emerged as a critical element in combating environmental pollution in the interest of sustainability. This chapter discusses key innovations in technologies that fall in the following categories: biochemical, thermochemical, mechanical and physical and advanced waste treatment processes. It presents all the categories in detail, elaborating on the underlying principles, applications, and recent advances. Finally, the chapter summarizes the challenges and future prospects of waste valorization technologies. 1. Introduction The ever-increasing population and rapid industrialization around the world have increased the generation of waste considerably and have become a serious threat to the environment.Landfills are filling up very fast, and emissions from waste disposal sites are contaminating air, water, and soil.

To tackle these challenges, waste valorization has emerged as an important strategy. Waste valorization is the process of converting waste materials into valuable resources, such as energy, fuels, chemicals, and materials, thus reducing environmental impact and promoting a circular economy. Technological advances in waste valorization have led to a wide scope of providing innovative solutions for a whole range of waste streams. This chapter provides an all-inclusive review of significant technological innovations in waste valorization, categorized into biochemical processes and thermochemical processes. Biochemical Processes. Composting, Anaerobic Digestion, Fermentation, and Bioenergy Production: Thermochemical Processes: Pyrolysis, Gasification, Incineration, and Biochar Production:.

Mechanical and Physical Processes: Milling, pressing, extraction techniques, and producing fiber-based products and bio-materials.

Advanced Waste Treatment Technologies: Biotechnology, nanotechnology applications, and smart waste management systems.

Biochemical Processes

Biochemical processes rely on biological agents mainly, on microorganisms. The transforming of organic wastes into product materials is usually characterized with relatively moderate operating conditions and huge opportunities of bio-based products' ranges.

Composting

Composting is a natural process involving the aerobic decomposition of organic matter, for example, food scraps, yard waste, and agricultural residues, into a nutrient-rich soil amendment known as compost (USEPA, 2023). It involves a controlled environment optimized to enhance microbial activity resulting in the stabilization and sanitization of the organic material (Zhang et al., 2018).

Process Overview

The composting process is usually composed of several steps:

Preparation: The organic waste is shredded or chopped to increase surface area and facilitate microbial activity.

Mesophilic Phase: Microorganisms thrive at moderate temperatures (30-40°C), breaking down easily degradable organic matter.

Thermophilic Phase: As microbial activity increases, the temperature rises to 55-65°C, killing pathogens and accelerating the decomposition process.

Curing and Maturation: The compost is left to cool and mature during which the microbial activity reduces and the compost stabilizes.

Uses

Compost is widely used as a soil amendment for improving the structure of soil, increasing water retention and nutrient availability for plant growth. It can also be used for landscaping, erosion control, and greenhouse cultivation.

Benefits

- Reduces the volume of organic waste
- Produces a valuable soil amendment
- Reduces greenhouse gas emissions from landfills
- Promotes sustainable agriculture

Challenges

- Requires careful management of moisture content, aeration, and temperature.
- Can produce unpleasant odors if not managed properly.
- May require significant land area for composting facilities.

Anaerobic Digestion

This process occurs by breaking down organic matter without oxygen, resulting in the production of biogas. Biogas consists mainly of methane and carbon dioxide (Appels et al., 2011). Biogas may be used as renewable energy to generate electricity or as fuel for heat purposes, and it can be converted into a vehicle fuel as well (Sari et al., 2014).

Overview of Process

There are generally four stages that usually comprise the anaerobic digestion process:

Hydrolysis: Complex organic matter is broken down into simpler molecules, such as sugars and amino acids.

Acidogenesis: Acid-forming bacteria convert simple sugars into organic acids, such as acetic acid and propionic acid.

Acetogenesis: Acetogenic bacteria convert organic acids into acetic acid, hydrogen, and carbon dioxide.

Methanogenesis: Methanogenic bacteria convert acetic acid, hydrogen, and carbon dioxide into methane and carbon dioxide. 2.2.2 Applications

Anaerobic digestion can be utilized for a wide variety of organic waste streams, including food waste, animal manure, and agricultural residues. It is also used in wastewater treatment and sludge stabilization.

Advantages

- Produces renewable energy in the form of biogas
- Reduces greenhouse gas emissions from landfills
- Reduces the volume and toxicity of organic waste
- Can be used for wastewater treatment

Challenges

Needs to have a controlled environment to preserve anaerobic conditions.

Is susceptible to inhibition by specific types of contaminants, for example heavy metals.

Will typically require large amounts of capital to construct facilities.

Fermentation

Fermentation is a metabolic process where microorganisms convert carbohydrates into organic acids, alcohols, or other chemicals (Soccol et al., 2010). This process is used in the production of various bio-based products, including bioethanol, lactic acid, and other value-added chemicals from organic waste streams (Mussatto et al., 2011). 2.3.1 Process Overview

Fermentation usually involves the following steps:

Substrate Preparation: The organic waste is pretreated to make it suitable for microbial conversion.

Inoculation: This step involves the inoculation of microorganisms, which include bacteria or fungi, in a substrate.

Fermentation: The microorganisms are able to metabolize the substrate, thereby resulting in the production of desirable products.

Product Recovery: The desirable products are separated from the broth containing the fermentation.

Applications

Bio-based products can be produced in fermentations from organic waste. Examples include:

- Bioethanol: It is derived from lignocellulosic biomass, such as agricultural residues and wood wastes.
- Lactic Acid: Obtained from diversified carbohydrate sources, including whey, corn starch, sugar beet molasses, among others.

- Other Chemicals: Prepared from organic wastes streams including butanol, acetone, and other organic acids.

Benefits

- Utilizes renewable resources
- Produces value-added chemicals and materials
- Reduces dependency on fossil fuels
- Can be compatible with other technologies for waste valorization.

Challenges

- Sensitive to shifts in the environment.
- Must in most instances process optimization so as to be able to achieve high yield.
- Can be vulnerable to contamination by unwanted microorganisms.

Bioenergy Production

Bioenergy production is the process of converting organic waste into energy in different forms, including biogas, bioethanol, and biodiesel. These biofuels can replace fossil fuels, thereby reducing greenhouse gas emissions and increasing energy security.

Biogas Production

Biogas is generated through anaerobic digestion of organic waste as explained above. It has a wide range of applications, including:

- Electricity generation
- Heat production
- Vehicle fuel

Bioethanol Production

Bioethanol is produced by fermenting sugars from various sources, such as:

- Sugarcane
- Corn
- Lignocellulosic biomass

Bioethanol can be mixed with gasoline to increase fuel efficiency and decrease emissions.

Biodiesel Production

Biodiesel is produced by the transesterification of vegetable oils or animal fats. It can be used directly in diesel engines in place of diesel fuel.

Benefits

Decrease reliance on fossil fuels

Reduces greenhouse gas emissions

Enhances energy security

Enhances rural development

Challenges

This technology can compete with food production for land and other resources.

Significant infrastructural investment may be necessary.

It can be exposed to feedstock supply fluctuation and price swings.

Thermochemical Processes

The thermal processing of organic waste converts this into energy or valuable products using the process of thermal treatment. Such processes can convert almost any waste stream to another product, including MSW, industrial waste, and agricultural residues.

Pyrolysis

Pyrolysis involves the thermochemical decomposition process of organic materials in the absence of oxygen, leading to the formation of bio-oil, biochar, and syngas, as concluded by Mohan et al. (2006). Bio-oil can be further upgraded for transportation fuels or chemicals whereas biochar can be further used as a soil amendment or as an adsorbent, for example, as concluded by Lehmann et al. (2011).

Process Overview

Pyrolysis is generally a process where organic waste is heated to a high temperature of 400-800 C in an oxygen-free environment. It can be performed in a number of reactors, which include:

- Fixed-bed reactors
- Fluidized-bed reactors
- Rotary kilns

Products

- Bio-oil: A liquid product that can be used as a fuel or feedstock for chemical production.
- Biochar: A solid carbon-rich product that can be used as a soil amendment or adsorbent.
- Syngas: A gaseous product that can be used for electricity generation or chemical production.

Waste Valorization: Turning Challenges into Opportunities

Trash Trouble: Our Landfills Are Rapidly Reaching Capacity - Food Cycle Science

Applications

Pyrolysis is applied to the following variety of waste streams:

- Municipal solid waste
- Industrial waste
- Agricultural residues
- Tires
- Plastics

Advantage

- It can be utilized for the conversion of diverse streams of wastes into usable products.
- Relatively clean fuels and chemicals are obtained.
- Biochar is another valuable product that can be generated.
- Decreases the volume and toxicity of the waste.

Challenges

- Needs high temperatures and sophisticated equipment.
- Producing complicated mixtures of products calls for further processing.
- May involve a large input of energy to heat the process.

Gasification

Gasification is a thermochemical process by which organic matter is transformed into a flammable gas mixture, referred to as syngas, under partial oxidation at high temperatures (Brandt et al., 2009). Syngas can be used for electricity generation, heat, or as a feedstock in chemical synthesis (Kurkijärvi et al., 2007).

Process Overview

The gasification process usually includes the following steps:

- Drying: The organic waste is dried to reduce the moisture content.
- Pyrolysis: The dried waste is heated in the absence of oxygen to produce volatile gases.
- Oxidation: The volatile gases are partially oxidized with air or oxygen to produce syngas.

Products

Syngas: A mixture of carbon monoxide, hydrogen, carbon dioxide, and methane.

Applications

Syngas can be used for:

- Electricity generation
- Heat production
- Chemical synthesis (e.g., methanol, ammonia)
- Fuel cells

Advantages

- Can be used to convert a wide range of waste streams into a versatile energy source.
- Produces a relatively clean fuel.
- Can be integrated with other energy systems.

Challenges

- Requires high temperatures and specialized equipment.
- Can produce tar and other contaminants that require removal.
- May require significant energy input for the gasification process.

Incineration

Incineration is a thermal high-temperature process where the combustion of waste materials is employed to reduce the volume with the production of heat or electricity (USEPA, 2021). Despite the efficiency of incineration in volume reduction of waste, proper air pollution control measures should be exercised to avoid emissions of pollutants harmful to human health and the environment (Wärnhammar et al., 2003).

Process Overview

Incineration typically means burning of the waste material at very high temperatures of 800 to 1100 C within a controlled environment. It is possible to make the steam to generate electricity, or in other industrial purposes by burning the produced heat.

Benefits

- Significant volume reduction of the wastes
- Ability to generate power and/or heat
- Capability for recovery of metals in waste

Drawbacks

- The risk to air: dioxin, furans, heavy metals
- Air pollution controls must be very tight.
- Public perception issues on possible environmental and health effects.

Biochar Production

Biochar is a carbon-rich solid product obtained by the pyrolysis of biomass. It can be used as a soil amendment that improves soil fertility, water retention, and carbon sequestration (Lehmann et al., 2011).

Process Overview

Biochar is usually produced through slow pyrolysis of biomass in the absence of oxygen at temperatures between 300-700 C.

Applications

Biochar can be used for:

- Soil amendment that enhances soil fertility and its water retention capacity.
- Carbon sequestration.
- Water and air purification.

Animal feed

Benefits

- Improves soil quality and agricultural productivity.
- Sequesters carbon and reduces climate change effects.
- Can be used in many environmental applications.

Problems

- Needs proper management to prevent detrimental environmental effects.
- May require further research to optimize the production and use of biochar.

Mechanical and physical processes are those in which separation, sorting, and transformation of waste materials into valuable products are done by physical means. Such processes are normally used as preliminary steps within the valorization chains in order to recover valuable materials and prepare the waste streams for further processing.

Milling, Pressing, and Extraction Techniques

- Milling, pressing, and extraction techniques are used in order to separate and recover valuable materials from waste streams.
- **Milling:** Breaks the solid waste into smaller sizes, which makes further processing easier (Kumar et al., 2013).

Pressing: Compresses the material to decrease the volume and remove the liquids (Kamm et al., 2006).

Extraction Techniques: Such as solvent extraction and supercritical fluid extraction, are used to extract certain fractions from the waste materials (Chemat et al., 2012).

Fiber-Based Products and Bio-materials Production

Fiber-based products, such as paper and cardboard, can be recycled and reused multiple times (European Commission, 2018). Additionally, innovative bio-materials, such as bioplastics and biocomposites, are being developed from renewable resources, offering sustainable alternatives to conventional petroleum-based materials (Auras et al., 2011).

Advanced Waste Treatment Technologies

Advanced waste treatment technologies apply the most advanced scientific principles to upgrade waste valorization processes. The technologies seem to provide exciting opportunities to enhance the efficiency of waste treatment and generate new valorization routes.

Biotechnology and Nanotechnology Applications

Biotechnology: This refers to the application of microorganisms or enzymes to degrade or transform waste materials (Banerjee et al., 2017).

Nanotechnology: Allows for the development of new materials and processes in waste treatment, such as nanofiltration membranes and nano-catalysts (Kumar et al., 2018).

Smart Waste Management Systems

Smart waste management systems integrate information and communication technologies, including sensors, GPS, and data analytics to optimize waste collection, transportation, and processing (Goyal et al., 2017). These systems allow real-time monitoring of waste streams, predictive maintenance of equipment, and resource allocation that leads to enhanced waste management outcomes (Caputo et al., 2019).

Challenges and Future Perspectives

Even though waste valorization technologies have experienced remarkable development, there are still a number of challenges:

- Economic viability: Several waste valorization technologies require a high level of capital investment and cannot be economically viable in every context.
- Technological constraint: Some technologies may not be efficient, scalable, or applicable to certain streams of waste.

- Public acceptance: Public concerns over possible environmental and health impacts associated with certain waste valorization technologies can prevent their use.
- Integration and coordination: Effective integration and coordination between different waste valorization technologies and stakeholders are crucial for successful implementation.

Future research and development efforts should focus on

Developing more cost-effective and efficient waste valorization technologies.

Improving the sustainability and environmental performance of existing technologies.

Addressing public concerns and promoting public acceptance of waste valorization initiatives.

Fostering collaboration between researchers, industry, and policymakers to accelerate the development and deployment of innovative waste valorization solutions.

Technological innovations are playing a pivotal role in advancing waste valorization practices, transforming waste from a liability into a valuable resource. Integration of biochemical, thermochemical, mechanical and physical, and advanced waste treatment technologies will help create a more sustainable and circular economy where waste is minimized and resources are maximized. Continued research and development in these areas are essential to address the growing challenges of waste management and promote a more sustainable future.

CHAPTER 5

VALORIZATION OF FRUITS AND VEGETABLE WASTE BY MICROENCAPSULATION

Fruits and vegetables are a vital component of the human diet as they provide essential nutrients, vitamins, and minerals. There is a boom in demand for fruits, vegetables, and their processed goods in the worldwide market. The food industry, particularly in FV processing, is generating waste due to an increase in production rates and inappropriate handling techniques. The waste is the major challenge faced by the global food industry. "According to the Food and Agriculture Organization, around a third of all food produced worldwide is lost or wasted (~1.3 billion metric tons)" (Hussain et al., 2020). Nearly half, i.e., up to 60% of horticultural produce (fruits, vegetables, and root crops), goes to waste worldwide. Moreover, between 30–50% of the growing FV have been lost even before they have reached to be consumed (Bisht et al., 2023). Nowadays, the growing trend that has been observed in consumers is about their concern regarding diet and the health impact of the consumed product. Thus, demand for more natural and safe ingredients is increasing, leading to the opening of an important research area that minimizes waste and also supports the concept of zero waste (Sharma et al., 2021). In the present scenario, landfills are the major problem that needs to be addressed. This problem can be solved by a sustainable approach towards the utilization and organization of industrial food waste. FVW also contains a health-promoting potential, which includes well-established sources of dietary fiber, pectin, polysaccharides, and several BC (Jiménez-Moreno et al., 2020). Innovative technologies like enzyme-assisted extraction (EAE), ultrasound-assisted extraction (UAE), supercritical fluid extraction (SFE), microwave-assisted extraction (MAE), pulsed electric field-assisted extraction (PEFE), and pressurized liquid extraction

(PLE) are used for the conversion of FVW into various value-added BC such as anthocyanins, flavonoids, carotenoids, dietary fibers, antioxidants, phenols, pectin, and oils (Râp *et al.*, 2024).

The Proper FVW utilization and management lead to a strengthen the economy, which can be increases the value of the waste material. Valorization is sustainable conversion by maximum use that leads to a value addition of the waste to enhance the economic value. (Râp *et al.*, 2024). Valorization of FVW can helps in resource conservation, product development, and waste utilization by reducing the cost.

A circular economy can achieve effectively turning waste into wealth through the valorization of FVW (Cassani *et al.*, 2022; Sharma *et al.*, 2021). At present consumers are more focusing on natural, healthy, or organic food that is safe for health. The increase an interest among consumers in FFs has leads to an increase demand for natural foods. The demand for FFs worldwide was estimated to be around $100 billion in 2013 (Borah *et al.*, 2023). The incorporation of certain health-promoting components is necessary for the production of FFs. Diabetes, cardiovascular illnesses, and cancer are the leading causes of death globally. The use of various FFs in the day-to-day diet could help to prevent certain disorders.

Microencapsulation is a technique that involves encapsulating active ingredients within a protective coating (Comunian et al., 2021). Because of its potential to enhance the stability, bioavailability (BA), and controlled release of encapsulated compounds, this approach has attracted a lot of attention lately (Cassani *et al.*, 2022; Cazón & Sanches Silva, 2024). In the context of FVW valorization, microencapsulation can be used to preserve BC and enhance their functional properties (Marcillo-Parra et al., 2021; Pattnaik *et al.*, 2021). The method can meet the demands of natural pigments, BC, essential oils, enzymes, vitamins, and minerals. This imparts rich biotherapeutic potential as well as contributes to the development of FFs (Pedro et al., 2024; Yadav *et al.*, 2024). Due to its beneficial properties, microencapsulation is extensively used by the food and pharmaceutical industries. The functional qualities of microencapsulates facilitate its transportation and easier handling of BC and also contribute to preventing or minimizing the release of the core material until it is encountered and also contribute to a decrease in the oxidation reaction of the core material when in contact with various environmental factors such as air, light, heat, and moisture (Qin, 2016).

The present review paper on the valorization of fruit and vegetable waste (FVW) aims to investigate unexplored areas of this topic, namely the incorporation of microencapsulation technology, which distinguishes present study from previous research. Although FVW valorization has been the subject of numerous published studies that concentrate on extraction techniques and applications in food products, these studies frequently ignore the stabilization issues related to bioactive compounds (BC) and the function of sophisticated encapsulation techniques in maintaining their functional properties. In addition to highlighting effective preservation techniques, the current review improves BC bioavailability, controlled release, and stability by utilizing microencapsulation techniques like spray drying and freeze drying. These factors are crucial for expanding the usefulness of FVW in a variety of applications. The present work is novel since it thoroughly examines encapsulation as a means of producing FVW-derived molecules in large quantities for the food sector. Present perspective matches a circular economy approach, meeting the environmental as well as consumer demand for natural, health-promoting components. Furthermore, present review offers future opportunities for encapsulating developments, including the use of biopolymers and possible uses in functional meals and nutraceuticals, which could provide important insights into creating more valued, sustainable products from agricultural waste.

Nutritional and Functional Significance of fruits and vegetable waste

FV are essential for both physical and mental health due to the presence of various phytochemicals. These phytochemicals, like dietary fibers, pigments, phenolic compounds, and micronutrients, have been shown to lower the risk of various health related issues such as cancer, cardiovascular diseases, hypertension, diabetes, overweight, and obesity. Moreover, the neuroprotective phytochemicals found in FV can enhance cognitive ability (Kakkar et al., 2021). "The majority of people eat the flesh or pulp of the most FV, but studies have shown that the peels, seeds, and other parts of FV that are not consumed often contain huge amounts of phytochemicals and essential minerals" (Kainat et al., 2022). For example, citrus peels, such as lemon, grape, and orange peels, as well as the seeds of mangoes, jackfruits, avocados, and longans, contain quantities of phenolics more than 15% greater than those in the fruit pulp (Sagar et al., 2018). The BC found in different FV, along with their associated therapeutic properties, are depicted in table 1 and table 2.

Table 1. Bioactive Compounds, methods of extraction, therapeutic properties and Potential Applications of different Fruits waste

Fruits	Type of waste	Methods of Extraction	Bioactive compound	Therapeutic properties	Potential applications	References
Black currant	Pomace	UAE	Anthocyanins, quercetin, myricetin	Antioxidants	FFs, prebiotic cosmetics	(Farooque et al., 2018; Petrov Ivankovi et al., 2024; Kruszewski & Boselli, 2024; Maryam Adilah et al., 2024)
Red grape	Pomace	MAE, UAE	Malvidin, quercetin	Anti-inflammatory, anti-angiogenic	Vermi-composting, FF	(Drosou et al., 2015; Caponio et al., 2023)
Banana	Peel	MAE	Hydroxycinnamic acids, catecholamines, isoquercitrin, myricetin, naringenin, rutin	Anti-hypertensive, antidiabetic, anticancer, antimicrobial	Food packaging, bioplastic film, edible food wrapper, buns, bread, pasta, confectionery products, natural preservative, noodles, heavy metal absorber, bio-ethanol production, bio methane production	(Zaini et al., 2022; Farooq et al., 2021; Putra et al., 2022)
Dragon fruit	Peel	UAE, MAE	Betanin, phyllocactin	Antioxidants, antimicrobials	Food preservation, food packaging	(Thaiudom et al., 2021; Andrade et al., 2024; Carrera et al., 2021; Thirugnanas-ambandham & Sivakumar, 2017)
Pomeg ranate	Peel	UAE	Punicalagin	Anti-inflammatory, antimicrobial	Food coatings, edible films, goat meatballs, chicken meat patties, beef burger, muffin cakes, cupcakes, biscuits, curd, fermented milk beverage, edible oil, nutraceuticals	(Siddiqui et al., 2024; Kandylis & Kokki-nomag oulos, 2020; Singh et al., 2023; Rakshit et al., 2020)

Fruits	Type of waste	Methods of Extraction	Bioactive compound	Therapeutic properties	Potential applications	References
Water melon	Rind	EAE	Ferulic acid, sinapic acid	Anti-hypertensive	Biosorbent	(Mushtaq et al., 2015; Bhattacharjee et al., 2020)
Apple	Pomace	UAE, MAE, EAE, PEFE	Dihydrochalcones, anthocyanins, flavonols	Antioxidant, anti-inflammatory	Dairy, confectionary, meat products, Bread, cakes, muffins, cookies, crackers, extruded food products, alcoholic beverages	(Zaky et al., 2024; -Kauser et al., 2023; -Lyu et al., 2020; Wang et al., 2023)
Orange	Peel	UAE, MAE, supercritical CO2 extraction, (SC-CO2)	Hesperidin, naringin, ferulic acid	Neuro protective, antioxidant, anti-inflammatory	Confectionery, bakery products, beverages	(Duysak & Kılıç, 2024; Tahir et al., 2023; -Irina et al., 2017)
Jackfruit	Peel	Sequential microwave-ultrasonic-assisted extraction	Chlorophorin, prenylgenistein, procyanidin B, feruloylglucoside, caffeoylglucoside	Anti-diabetic, antioxidant, antimicrobial, anti-inflammatory, anti-tumor	Antimicrobial packaging films, biodegradable plastics, bionan oparticles	(Brahma & Ray, 2022; Ly & Sothornvit, 2024)

Fruits	Type of waste	Methods of Extraction	Bioactive compound	Therapeutic properties	Potential applications	References
Mango	Peel	Subcritical water extraction (SWE), SFE	Protocatechuic acid, gallic acid, quercetin, mangiferin, -carotene, syringic acid violaxanthin, luteine	Antiprolife-rative anticancer, antidiabetic anti-inflammatory, antibacterial	Nutraceutical, food supplements, additives, beverages, gelling agent, emulsifier, thickener, noodles, sponge cakes, bread, biscuits	(Serna-Cock et al., 2016; Umamahesh et al., 2020; Wongkaew et al., 2021; Thambi et al., 2016; Tunchaiyaphum et al., 2013, Souza et al., 2019)
Pineapple	Peel	UAE, MAE	Catechin, ferulic acid, gallic acid, epicatechin	Antifungal, anti-inflammatory, antioxidant, anticancer	Cracker, vinegar, cake, cereal bar, yogurt, muffin, cookie, pectin	(Mala et al., 2024; Mala et al., 2021; Polania et al., 2023)
Jackfruit	Rind, seed	SWE, UAE, MAE, SFE	Gallic acid, tannic acid, ferulic acid, catechin, caffeoylglucoside, dihydromyricetin, chlorophorin, prenylgenistein	Anti-tumor, anti-inflammatory, antioxidant antidiabetic	Muffins, breads, biscuits, cookies, meat analogue, cocoa substitute, jelly, bun, pasta, tortilla, noodles, chips, diet chocolate, ice cream, bread spread, spaghetti	(Ramya et al., 2020; Aswin et al., 2022; Alibekov et al., 2023; Brahma & Ray, 2022)
Grapes	Seed extract, pomace	Accelerated solvent extraction, SFE	Isoflavones, procyanidins, flavanols, stilbenes catechin, epicatechin-3-O-gallate	Antioxidant, antimicrobial, antiviral, anti-inflammatory	Natural antioxidants, natural preservatives, food fungicide, food packaging film, pancakes, bread, extruded cereals, biscuits, pasta, meat products	(Chen et al., 2020; Zhou et al., 2022; Fontana et al., 2013)

Table 2. Bioactive Compounds, methods of extraction, therapeutic properties and Potential Applications of different vegetables waste

Vegetables	Type of waste	Methods of Extraction	Bioactive compound	Therapeutic properties	Potential applications	References
Broccoli	Leaves, stalk	MAE	Glucosinolates, isothiocyanates, chlorogenic acid, sinapic acid, sulforaphane, kaempferol	Anti-cancer, antioxidant, antibacterial, anti-inflammatory	FF, food additive, craft beers, gluten-free breads, sponge cakes, pasta, sauces, pickles	(Nartea, 2022; Yan et al., 2023; Salas-Millán et al., 2024; Shinali et al., 2024; Villaño et al., 2023; García, 2022)
Eggplant	Peel	UAE, EAE	Anthocyanins, malvidin, peonidin, cyanidin	Anticancer, antimicrobial, antioxidant	Food colorant, antioxidant agent, pharmaceutical products, pullulan, chocolate, jelly, ice cream, candy, jelly crystals, custard, antioxidants, pastry cream	(Karimi et al., 2021; Ferarsa et al., 2018; Amulya & ul Islam, 2023)
Onion	Peel	SFE, EAE, CSE	Quercetin 3,4′-diglycoside, kaempferol, protocatechuic acid, isorhamnetin, anthocyanins, tannins, vanillic acid, ferulic acid, quercetin, p-coumaric acid, epicatechin, morin	Antidiabetic, neuroprotective, cardioprotective, antiobesity, erectile dysfunction properties	Antioxidants, food colorant, substrate for enzyme production, biodegradable packaging materials, jellies, sugar-enrobed confectionery, fat-based coatings, foamed commodities, gluten-free bread, meat products, food packaging, preservative	(Kumar et al., 2022a; Kumar et al., 2022b; Shabir et al., 2022; Constantin et al., 2021)

Vegetables	Type of waste	Methods of Extraction	Bioactive compound	Therapeutic properties	Potential applications	References
Beetroot	Leaves	SC-CO2, PLE	Apigenin, vitexin, rutin, quercetin, kaempferol, red betacyanins, yellow betaxanthins	Antibacterial, antiquorum sensing, antioxidant, antipathogenic agents	Food colorants, antioxidant	(Chaari et al., 2023; Pellegrini & Ponce, 2020; Mella et al., 2024; Goyeneche et al., 2020; Lasta et al., 2019)
Tomato	Peel	High-hydro-static pressure extraction, SFE	Carotenoids, lycopene	Anticancer, antioxidant	Food coloring, cookies, ice cream	(Jiménez Bolaño et al., 2024; Pellicanò et al., 2019; Grassino et al., 2020)
Carrot	Pom-ace	CSE, MAE, UAE	Zeaxanthin, Lutein	Antioxidant, antidiabetic	Cookies, pasta, biscuits, bread, meat processing, dairy processing, antioxidant, nutraceuticals	(Šeregelj et al., 2020; Kaur et al., 2020; Stamatovska et al., 2024; Nazar et al., 2023)
Bottle gourd	Peel	UAE	Cucurbitacins, triterpenoid, flavone C-glyco-sides	Anticancer, anti-diabetic, cardioprotective, anti-inflammatory properties.	FF, adsorbents for the removal of reactive red 195-A and reactive blue 222 dye, instant kheer, pasta, cookies, biscuits	(Palamthodi & Lele, 2016; Sharma et al., 2023; Abrol et al., 2023; Lau et al., 2021)
Radish	Peel, seed	UAE	Isothiocyanates, anthocyanins, glucosinolates.	Antidiabetic, anti-inflammatory, immunoprotective, hepatoprotective, diuretic activities, anti-hyperlipidemic, anti-hypertensive, anti-atheros-clerotic.	Natural food colorant, traditional medicine	(Chihoub et al., 2019; Ibrahim et al., 2024; Matus-Castillo et al., 2022; Sham et al., 2013; Stevanato & da Silva, 2019)

Vegetables	Type of waste	Methods of Extraction	Bioactive compound	Therapeutic properties	Potential applications	References
Pumpkin	Peel, seeds, pulp	SWE, SFE, MAE, UAE	Lutein, phytosterols, zeaxanthin	Cardioprotective, antioxidant, anti-inflammatory, anticancer, antiaging	Bread, muffins, biscuits, chicken burgers, beef patties, yogurt, ice cream, nutraceutical, dietary supplement	(Aziz et al., 2023; Gavril et al., 2024; Ezzat et al., 2022)
Pea	Pods	UAE	Tannins, 5-caffeoylquinic acid, terpenoids, hesperidin, epicatechin	Anticancer, antioxidant	FFs, nutraceutical, synthesis of biocolors, Snacks, dry soups, biodegradable films, cellulase production, natural color, prepare enzymes, pectin	(Abd El-Galil et al., 2021; Fatima et al., 2024; Mousa et al., 2021; Pinchao-Pinchao et al., 2019)
Capsicum	Peel, seed	PEFE, SFE, UAE, EAE, MAE	dihydrocapsa-icin, capsaicin, capsaicinoids	Antioxidants, anti-carcinogenic, anti-inflammatory, anti-diabetic, anti-obesity and anti-microbial.	Pharmaceutical products, FF	(Olguín-Rojas et al., 2024; Yasin et al., 2023; Echave et al., 2020)
Potatoes	Peel, fiber	Seque-ntial hydrothe-rmal extraction, PLE, MAE, UAE	Chlorogenic acid, protocatechuic acid, caffeic acid, catechin, kaempferol, rutin, quercetin	Anti-inflammatory, chemopreventive, antioxidant, antimicrobial	Pasta, cakes, bread, natural antioxidants, yogurt, food coating film, thickening or stabilizing agent, biscuits	(Jimenez-Champi et al., 2023; Xie et al., 2018; Rodríguez-Martínez et al., 2021)

Vegetables	Type of waste	Methods of Extraction	Bioactive compound	Therapeutic properties	Potential applications	References
Spinach	Leaves	PLE, SC-CO2	Apigenin, quercetin, kaempferol,	Antiallergic, anticancer, anti inflammatory, antioxidant	Cheese, meat products, antimicrobial agent, antioxidant, fried food, cake	(El-Sayed, 2020; Salehi et al., 2019; Koç et al., 2019; Singh et al., 2018; Jaime et al., 2015)
Okra	Seed, upper part (crown)	MAE	Quercetin	Antioxidant, antidiabetic	Ice cream, chocolate, mousses, chips, cookies, cakes, tortillas as an emulsifier, fat replacer, texture modifier, gelling agent, thickening agent, stabilizer, flavouring agent, emulsifier	(Dantas et al., 2021; Durazzo et al., 2018; Tufaro et al., 2022; Williams et al., 2023; Woumbo et al., 2022)
Garlic	Peel	High pressure extraction	Caffeic acid, ferulic acid, -coumaric acid, tellurium, organoselenium	Neuroprotective, hepatoprotective, antifungal, antioxidant, anticancer, antidiabetic, cardioprotective	Food packaging films, natural preservative in wheat flour, flour-based food products	(Chaudhary et al., 2021; Hernández Montesinos et al., 2023; Hitlamani & Inamdar, 2024; Shen et al., 2024; Carreón-Delgado et al., 2023; dos Santos et al., 2022; Azmat et al., 2023)

Microencapsulation of Bioactive Compounds from Fruits and Vegetable Waste by Using Spray drying

Lim *et al.* (2019) studied spray drying (SD) technology to microencapsulate phenolic compounds extracted from waste mango seed kernels (MSK). MSK are abundant in phenolic compounds, which have promising applications in the nutraceutical sector and in the development of FFs due to their numerous health benefits. A laboratory-scale spray dryer operated at three distinct temperatures (130, 150, and 170 C) and three different encapsulating agents (EA), soluble starch (ST), gum arabic (GA), and maltodextrin (MD), at a working concentration of 10% (w/w). The study indicated that MD yielded the highest encapsulation rate, signifying a strong interaction between the bioactive portion of mango seed kernel extract (MSKE) and MD in comparison to the other EAs. Moreover, the water solubility index for powders made with MD and GA was higher than that for ST. The SD of MSKE produced powder with TPC levels ranging from 77 to 127 mg per gram of MSKE powder and antioxidant activity (AA) ranging from 86,740 to 217,700 mmol Trolox per 100 grams of MSKE powder.

Olt *et al.* (2021) evaluated the encapsulation efficiency of BC encapsulated from Tannat grape peel extract as a natural color and antioxidant in yogurt. A SD technique with NHWPI and IN, and HWPI and IN, both in a 3:1 ratio as EA, were used in the two encapsulation systems. The two encapsulation systems were spray-dried at 180 C to obtain powdered samples. The encapsulated extract was then added to yogurt formulations to determine its antioxidant activity and hue value. The findings showed that the NHWPI and IN formulation had the highest encapsulation efficiency (EE) percentage at 29.65%, whereas that of HWPI and IN was 12.81%. NHWPI-encapsulated extract-fortified yogurt reported the greater antioxidant capacity than the base formulation and the heat-treated extract, which indicated a protective effect of the encapsulation on anthocyanin degradation. The extract also imparted purple color to the yogurt, revealing potential as a natural colorant or as a functional ingredient in yogurt.

Fiorentini *et al.* (2021) studied the concept of commercial citrus peel extract(CCPE) incorporation into the polymeric matrix for development of active food packaging. Present study use a SD process where kleptose/β-cyclodextrins were used as EAs. Then, encapsulated CCPE films were developed in PLA and PHB films to be used based on their antimicrobial and antioxidant activities against *Escherichia coli* and *Staphylococcus aureus*. The encapsulation process reduced the TPC by 20–25%, while AA remained unaffected. The antimicrobial activity of films incorporated with CCPE was quantified in terms of growth inhibition percentages as 60 and 30 for *Staphylococcus aureus* and *Escherichia coli*, respectively. Citrus peel extracts encapsulated offered promising bio-based additive to enhance the antimicrobial activities in active food packaging.

Ramírez Damián *et al.* (2022) encapsulated the BC extracted from red banana peels by using SD by a matrix formed by using soy protein isolate (SPI) and GA as EAs. The encapsulation efficiency (EE) of TPC was reported as 98.3% with excellent retention efficiency of 98.8%. In vitro digestion was carried out to evaluate the bioaccessibility (BI) of the encapsulated extract. The results indicate that the encapsulation process protects BC against degradation during digestion. Higher retention of antioxidant capacity and BC was observed for the encapsulated extract as compared to the unencapsulated extract. The results may point to the possibility of microencapsulated red banana peel extract being a valuable ingredient in developing FF with increasing BI of BC.

Bastos *et al.* (2024) encapsulate strawberry extract and its principal BC like quercetin using SD technology. The extract microparticles exhibited one of the most promising formulations to be used for the incorporation of strawberry plant extract in cosmetic, nutraceutical, and food products, presenting high EE and a stable and well-defined release profile. The study used various biopolymer-based EAs, including single biopolymers (alginate, carrageenan, or pectin) and binary or ternary mixtures. The concentration of the EA solutions was maintained at 1% (w/v), while the final formulation contained 2% (w/w) of the active compound. The SD process was optimized, and its parameters were air pressure (5-6 bar), aspiration rate (36 m /h), solution feed flow (4 mL/min), inlet air temperature (115 C), and the outlet air temperature varied between 47 and 61 C. The resulting microparticles showed high encapsulation efficiencies, ranging from 77 to 99%, and the binary mixtures containing alginate showed the highest efficiency of 100 ± 1% for quercetin. The microencapsulation of anthocyanin extract from purple cabbage used for stability of color in fermented milk drinks was carried out by Espinosa Alvarez *et al.* (2019). The extraction of anthocyanin was made through the microwave assisted extraction method and microencapsulation was carried out by SD method by using three different types of wall materials: GA, MD, and IN. The spray dryer was used under two different inlet air temperatures at 150 and 170 C. The optimal SD conditions were found to be an inlet and outlet air temperature of 150 C and 90 C, respectively, with a MD-anthocyanin concentrate ratio of 2:1, showing a maximum yield of 58.9%. Various concentrations of microencapsulated powder, ranging from 0.20 to 0.40 grams, were added to 25 mL of milk beverage; the 0.3 concentration showed the least variation in color with a color difference (AE) value of 3.57. Overall, the study concluded that SD is a promising method for encapsulating anthocynin extracted from purple cabbage for its application in fermented milk drinks.

Microencapsulation of bioactive compounds from fruits and vegetable waste by using freez drying

Elsebaie & Essa, (2018) studied the encapsulation polyphenol extracted from red onion peel and its application in cake as a functional ingredient. The study utilized freeze drying (FD) techniques with SPI, MD, and a complex of MD and SPI. The pilot-scale freeze dryer was used to freeze dry the three encapsulated systems at -40 to -30 C for 20 hours to obtain powder samples. The encapsulated extrac (3 g) was added to wheat flour 100 g while preparing the cake. Results indicated that the complex of MD and SPI demonstrated the highest EE of 94.30%, while that of MD and SPI was 91.50% and 89.83%. Freeze-drying proved to be an effective method for encapsulating the red onion fraction, achieving high encapsulation efficiencies of up to 89.83% across all three encapsulating shell substances. After baking, the total polyphenol content retained 51.30% of its original value, demonstrating the strong protective effect of the encapsulation materials. Cakes formulated with microencapsulated polyphenolic extracts exhibited superior sensory attributes (color, taste, aroma, texture, and overall acceptability) compared to those made with direct addition of extract.

Martinovi *et al.* (2024) work on the in vitro BI of microencapsulated grape pomace extract. The study utilized the freeze-drying method with three coatings: sodium alginate (SA), combination SA and gelatin (GEL), SA and GA. Samples were frozen at 80 C overnight, followed by freeze-drying for 24–48 hours at 0.250 mbar and then crushed into a powder. Results indicated that the complex of SA-GEL demonstrated the highest EE for all samples, which ranged from 97.02 to 98.30%, while that of SA-GA and SA was 90.71 to 93.40% and 79.79 to 84.29%. The release of TPC from microencapsulated grape pomace extracts was simulated in three stages: oral, gastric, and intestinal, using electrolyte solutions without enzymes to stimulate the conditions in the upper human digestive tract. The highest BI values were observed for microencapsulated SA-GEL powders, ranging from 47.1 to 70.9% for vanillic acid and 23.6 to 80.7% for ellagic acid, under in vitro assessment.

Pashazadeh *et al.* (2021) studied encapsulation of phenolic compounds extracted from maize waste by different encapsulation techniques using MD as a wall material. The study utilized encapsulation of the sample by three techniques: SD, FD, and microwave drying. The samples were first frozen with liquid nitrogen and then freeze-dried for 48 hrs. Microwave drying was conducted in a household microwave oven at a power output of 750 W for a duration of 10 minutes. The analysis revealed that among the drying methods tested, freeze-drying was the most efficient, maximizing the retention of catechin (5.58μg/g), vanilin (17.79μg/g), ρ-cumaric acid (112.83μg/g), salicylic acid (17.32μg/g), and protocatechuic acid (152.06μg/g). Freeze-drying proved to be the most effective technique for preserving the bioactive compounds, resulting in a retention of 98.85% TPC, 97.95% total flavonoids (TF), 70.39% DPPH (2,2-diphenyl-1-picrylhydrazyl) radical scavenging activity, and 82.45% FRAP (Ferric reducing antioxidant power) activity. The glass transition temperatures were determined to be 171.09 C, 143.40 C, and 148.25 C for microwave-dried, spray-dried, and freeze-dried samples. The microcapsules can be used further in the development of FF.

Fuentes *et al.* (2023) focused on encapsulating BC from Chilean papaya waste, utilizing both the seeds and skin. The study utilized FD techniques with different concentrations of MD as a wall material that was freeze-dried for 24 hr at 40 C. Results indicated that 20% MD demonstrated the highest EE for papaya seed extract, which ranged from 91.7%, while 10% MD showed the highest EE for papaya skin extract, which ranged from 90.7%. The highest amounts of phenolic and flavonoid compounds were recorded in extracts prepared by using MD20 and MD10 seed and skin extracts as compared to others. Moreover, they exhibited excellent antioxidant activity and significantly lowered intracellular Reactive oxygen species (ROS) values. Hence, encapsulated papaya extracts could probably provide health benefits and were applicable in food as well as pharmaceutical applications.

Alabidein *et al.* (2024) microencapsulate mandarin fruit peel, which is one of the novel sources to produce extracts rich with bioactive components. Mandarin peel extract (MPE) was mixed with MD and WP solution at 5% MD and 1% whey protein (WP) in equal proportion and microencapsulated using FD method. EE and encapsulation of the developed microcapsules were studied. The recorded encapsulation yield was 79.17%, and EE was determined as 55.26%. This application increases product durability by preserving the properties of MPE.

Mkhari *et al.* (2023) reported the ability of MD, GA, and their combinations to prepared functional beetroot waste extract powder (BWEP). The beetroot waste extract was mixed with 10% (w/v) GA and MD blends. Samples were later homogenized for 45 sec, frozen for 24 hr at 20 C, and then freeze-dried for 48 hr at -60 C and 0.01 millibar.the present study reported that the betalain encapsulation efficiency (BEE) of BWEP produced by GA and MT individually ranged from 82 to 88%. However, the combination of the two biopolymers resulted in a lower BEE, ranging from 64% to 70%. The study showed that using these biopolymers separately showed a promise to promote a food circular bioeconomy.

Spray Drying vs. Freeze Drying: Comparative study on Stability, Encapsulation Efficiency, Controlled Release, and Bioavailability Stability

Encapsulation can significantly enhance the stability of the encapsulated material (Stabrauskiene *et al.*, 2024). The process of encapsulation can significantly enhance the stability of the material under encapsulation. FD offers a more stable encapsulated product as compared to SD (Bano i *et al.*, 2023). The lyophilization process operates at a lower temperature, which preserves the natural pigments (Šeregelj et al., 2021). Freeze-dried products have lower exposure to oxygen, which results in a decrease in oxidative degradation of compounds and tends to have better oxidative stability of the product. At elevated temperatures during the spray-drying process, the oxidation of sensitive compounds, particularly polyphenols and carotenoids, is degraded to some extent (Drosou & Krokida, 2024; Šeregelj et al., 2021; Singh et al., 2024). The superior stability of heat sensitive compounds is offered by FD due to use of low temperature in this method while high temperature in SD results in thermal degradation of heat sensitive compounds (Drosou & Krokida, 2024; Šeregelj et al., 2021; Stabrauskiene et al., 2024). The encapsulate that was spray-dried displayed the lowest water activity, moisture content, and particle size, whereas the encapsulate produced through FD exhibited the highest levels of oxidative stability, hygroscopicity, and color characteristics (Singh et al., 2024). Both methods can enhance the stability of bioactive substances by shielding them from environmental influences such as light, oxygen, and temperature. The choice of wall material can significantly impact the stability of the encapsulated product. Maintaining the stability of both freeze-dried and spray-dried microcapsules requires appropriate storage conditions, including low temperature and low humidity (Drosou & Krokida, 2024; Šeregelj et al., 2021).

Encapsulation efficiency

Encapsulation efficiency plays a vital role in assessing the overall effectiveness of the microencapsulation process. It indicates the proportion of the active ingredient that is effectively contained within the carrier matrix. The EE of SD is higher than that of FD. (Perkovi et al., 2024) encapsulated the polyphenols extracted by grape pomace by SD method using WP as a wall material with an inlet air temperature of 180 C. The EE was 95.5%, with good antioxidant stability under these conditions. (Andrade et al., 2023) studied the encapsulation of phenolic compounds extracted from cashew apples using different concentrations of MD and IN as wall material by SD method with inlet and outlet temperatures (OT) of 140 C and 97°C. The EE was highest at 75%, and 100% MD concentrations were 96.8 and 96.25%. (Elsebaie & Essa, 2018) encapsulated the polyphenolic content extracted by red onion peels with FD method by different EAs. The sample with MD and SPI EA had higher EE (94.30%). Star fruit (*Averrhoa carambola*) pomace polyphenols was encapsulated by FD with MD as an EA. Freeze-dried encapsulates demonstrated a greater EE (78–97%) compared to spray-dried ones (63–79%); additionally, increasing the concentration of MD resulted in a higher content of core polyphenols. SD was conducted at an inlet temperature (IT) of 185°C and an OT of 88°C. For FD, frozen samples at 40°C were subjected to freeze-drying at 55°C for a duration of 24 hours (Buljeta et al., 2022). (Drosou & Krokida, 2024) encapsulated carotene using the SD method and freeze-drying method with WP and pullulan as a coating material. EE for SD was obtained at 53.78% and for FD at 63.69%.

The several above studies states that compounds that exhibit greater thermal stability, such as certain polyphenols like flavonoids, tend to have improved EE during SD. This is because they are less likely to break down under the high temperatures of the process. For instance, anthocyanins found in various fruit extracts have demonstrated better stability when encapsulated with high amounts of carriers such as MD (Barretto et al., 2020). Certain heat-sensitive compounds, such as certain vitamins (such as Vitamin C) and volatile oils, might degrade during the drying process, resulting in lower EE. This degradation of bioactive compounds is often due to exposure to high temperatures used for drying or prolonged exposure to air through the process (Rezvankhah *et al.*, 2020). The variation in value of EE with respect to SD depends on factors such as the thermal stability of the compound, carrier materials type concentration, and properties of the bioactive active compound itself as well as from conditions of the spray-drying operation (Acharya *et al.*, 2021; Mohammed *et al.*, 2020). The temperature was low and hence it would be relatively heat-sensitive, a factor that suits substances such as vitamins, polyphenols, and probiotics. In such cases, it exhibits improved retention during the freeze-drying process, and an immediate high level of EE is attained. Compounds exhibiting high volatility or very limited solubility in the carrier material may get a lesser EE value during FD. Further, the sublimation process may lead to evaporation or leakage of some BC if they are not well encapsulated. FD is particularly favored for compounds that are sensitive to heat and will offer higher EE for those prone to heat degradation. On the other hand, SD tends to get better EE when dealing with heat-stable compounds or where rapid drying and consistent droplet formation are critical (Acharya et al., 2021; Mohammed et al., 2020).

Controlled release and bioavailability

Encapsulation is one food preservation technique that increases the BA and stability of bioactive compounds found in FVW (Marcillo-Parra et al., 2021). The BA of BC shows the portion of a particular nutrient which could be digested, absorbed, and metabolized through regular metabolic pathways. Carotenoid has least BA obtained from natural sources. The microencapsulation of carotenoid was carried out, which improved the stability and regulated the release compounds observed in the human body. Several studies reported that encapsulation of carotenoids affect the BA. The in-vitro and in-vivo studies reported that microencapsulation of carotenoids enhances their BA (Eun et al., 2020). Controlled release is crucial in encapsulation as it allows for the effective utilization of the bioactive compounds (vitamins, antioxidants, and other nutrients) obtained from the sources. Encapsulation facilitates the controlled and targeted release of BC at a specific time. Thus, the controlled release process safeguards and preserves, ensuring the delivery of these bioactive elements to enhance their functionality at the appropriate moment (Bourbon et al., 2016). (Zhang et al., 2020) evaluated in vitro BI of anthocyanin extract from grape skin and microencapsulated anthocyanin extract of grape skin by both SD and freeze-drying methods. Relative effectiveness of anthocyanins in spray-dried products was found as 83% at the end of gastric digestion, whereas in freeze-dried products, it was reported to 70%. The anthocyanin extract of grape skin without encapsulation showed a continuous reduction of anthocyanin, which was found to be almost two-fold (45%) less than the spray-dried product. The retention of anthocyanins showed a notable decrease in the final stage of intestinal digestion; the retention of anthocyanins recorded were 1%, 15%, and 24.5% for free anthocyanin extract of grape skin, freeze-dried product, and spray-dried product, respectively. The BA flow was identified as spray-dried product >freeze-dried product > anthocyanin extract of grape skin. According to the results, powder microcapsules released anthocyanins more efficiently than anthocyanin extracts of grape skin during simulated gastric and intestinal digestion in vitro. This demonstrated that microencapsulation improved anthocyanin distribution under simulated gastrointestinal circumstances, enhancing BA and lowering anthocyanin degradation. Evaluated that the particle size directly affects the release of BC within the body. This is the reason for more BA of spray-dried product over freeze-dried product in the above data. Thus, the anthocyanin retention was higher in spray-dried products with smaller particle sizes as compared to freeze-dried products due to the large particle sizes. (El-Messery et al., 2020) studied the effect of different concentrations of krill oil (2–10% w/v) on vitro BI. The BI ranged from 20.0 and 83.0% for spray-dried encapsulated products and 32.7 and 72.4% for freeze-dried encapsulated products. The BI of spray-dried products was higher at 8 and 10% concentrations, and freeze-dried products showed the higher BI at concentrations of less than 8%. The variation in BI might be due to hygroscopicity, which mainly leads to lipid oxidation and affects the powder flowability. (Fredes et al., 2018) studied the BI after in vitro digestion of spray-dried maqui juice powder, freeze-dried maqui juice powder, and maqui juice. The BA of total anthocyanin in spray-dried powder was 44.1%, freeze-powder was 43.8%, and maqui juice was 35.2%. BI of specific anthocyanins of encapsulated powder was highest of cyanidin-3-sambubioside5-glucoside and cyanidin-3,5-diglucoside 76.5% for freeze dried product and 78.5% for spray dried product and lowest of delphinidin-3-glucoside 22.6% for freeze dried product and 20.8% for spray dried product. The result revealed that the primary differences between powders produced through spray-drying and freeze-drying were in their particle size, which influenced their solubility when reconstituted in water and morphology. The microencapsulation enhances their BA through several mechanisms, including increasing their stability against isomerization and degradation caused by various factors, improving their solubility, which leads to greater BI, and facilitating targeted release kinetics (Eun et al., 2020).The above studies reported that freeze-drying is typically more effective at preserving heat-sensitive substances, making it a better choice for ensuring the BA of sensitive molecules. While SD can produce particles with a greater surface area, potentially improving dissolution, freeze-dried formulations may still provide better BA for nutrients due to their enhanced stability and porous nature also spray dried product have better BA is some cases it is due to the particle size, core material used, thermo-stable compound and several factors.

Table 3. Encapsulation Techniques for Bioactive Compounds from fruits and vegetable Waste Materials: Specifications, Efficiency, and Applications

Technique	Waste material	Encapsu-lating agents	Conditions	Yield	References
SD	Cashew apple waste	MD (75 %) and IN (25 %)	IT (110ºC), flow rate (0.75L h 1), OT (76.20ºC)	Encapsulating efficiency (EE) (68.38%), retention of phenolic compounds (51.48%), TPC (61.85%) and ABTS (2,2′-azino-bis-(3-ethylbenzothiazoline-6 sulfonic acid) (48.65%)	(de Sena Andrade et al., 2023)
SD	Avocado oil waste	15% MD	Spray nozzle (0.7mm), IT (110ºC), OT (75ºC), flow rate (6 ml/min for 1L-1h)	EE (99.7%)	(Gonzales, 2024)
SD	Green okra leaves and seeds	MD and pectin (10:1)	IT (150ºC), OT (75ºC), feed rate (500 mL/h), spray nozzle (0.5 mm)	EE (73.84 to 90.6 %)	(Guebebia et al., 2023)
SD	Carrot waste	WP (71%) and IN (29%)	IT of (130ºC), OT (65ºC), spraying air flow rate (600 L/h), rate of liquid feed (8 mL/min)	EE (53.78%), TC (0.87 mg -carotene/100 g), BCBA (41.23 μmol TE/100 g)	(Šeregelj et al., 2021)
SD	Papaya seeds	Chitosan (0.5%)	Nozzle diameter (0.5 mm), feed flow (10 mL/min), vacuum (100%), IT (150ºC), OT (65, 68, and 70ºC)	Polyphenol content (97.6 mg *Gallic acid equivalent* /100 g sample) (98.9%), antioxidant capacity (413 mg TE/100 g sample)	(Hinostroza-Quiñonez et al., 2024)

Technique	Waste material	Encapsu-lating agents	Conditions	Yield	References
SD	Pine-apple peel	MD, IN, and GA (5% w/w)	Inlet air drying temperature (150ºC), inlet feed flow rate (3.7 mL/min), pressure (1.7 bar), drying air flow rate (47 m3 /h)	-	(Lourenço et al., 2020)
SD and high-speed homogenizer (HSH)	Apple peel	The first wall material- MD: WPC at ratio 8: 2 then PCEP was added (1: 20) Second wall material- GA mixture was added to the previous mixture at ratio (4: 6)	SD- IT (150ºC), OT (50ºC). HSH- (4000 rpm for 5 min)	Total phenolics content (TPC), Antioxidant activity (AA) and total flavonoids (TF) EE values of encapsulated PCEP were 86.85%, 83.69 and 85.47 mg catechin/kg for AA, PC and FC respectively when encapsulated by freeze drying	(El-Messery et al., 2019)
FD	Red Grape juice	Whey protein isolate (WPI) (12 % m/V) and chitosan (1.5 %)	Temperature (-42ºC), pressure (10 Pa), time (48hr)	EE (86.1%), anthocyanin mass fraction (cyanidin-3-O-glucoside equivalent) (1.4mg/g), flavonoid (catechin equivalents) (3.3mg/g), mass fractions (1.6mg/g)	(Mihalcea et al., 2020)
FD	Passion fruit peel	MD, waxy starch and GA (10% w/v)	Temperature (-60ºC), pressure (0.01 mbar), time (72hr)	EE (82.64 to 87.18%)	(Kobo et al., 2022)

Technique	Waste material	Encapsu-lating agents	Conditions	Yield	References
FD and ultrason-ication	Apple peel	The first wall material- MD: whey protein concentrate (WPC) at ratio 8: 2 then polyphenolic compounds extract powder (PCEP) was added (1: 20) Second wall material- GA mixture was added to the previous mixture at ratio (4: 6)	FD- drying temperature (-45ºC), drying duration (>48 hr), pressure (<0.120 mbar). Ultrasonication- (160 W power, 20 KHz frequency, 50% pulse)	EE values were 80.21%, 83.58% and 48.31% mg catechin/kg for AA, TPC and TF	(El-Messery et al., 2019)
FD, HSH	Onion peel	MD and casein (8:2, 6:4)	FD- (48 hr). HSH- (10000 rpm for 10 min)	EE (84.39–89.15%)	(Akdeniz et al., 2018)
FD, homogen-ization	Tomato peels	WPI and acacia gum: water (5:1) (w/w)	FD- temperature (-42 ºC), pressure (10 Pa), time (48 hr) Homogenization- 5000 g for 10 min	For two different samples- (AA (8.57 and 9.37 mMol Trolox Equivalent Antioxidant Capacity /g DW (dry weight)), -Carotene (22.49 and 19.93 mg/g DW), lycopene (10.83 and 10.49 mg/g DW), lycopene retention (48% and 29%))	(Mihalcea et al., 2021)

Technique	Waste material	Encapsu-lating agents	Conditions	Yield	References
Carrot waste	FD	Whey protein (WP) (100%)	Temperature (-40°C), time (48hr)	EE (63.69%), total carotenoids (TC) (1.31 mg -carotene/100 g), -carotene bleaching antioxidant capacities (BCBA) (41.23 μmol Trolox equivalents (TE)/100 g)	(Šeregelj et al., 2021)
Vacuum drying (ultrason-ication)	Apple pomace	MD	Wall to core ratio (7.5), the percentage of surfactant (4% w/w), time of sonication homogenization (2 min).	Loading efficiency (98%), polyphenol loading content (8.62%), MD-nanoparticles yield (75%)	(Saffarzadeh-Matin & Shahbazi, 2017)
Vacuum freeze drying and ultrasound homogenizer	Red dragon fruit peel	Betalains extraction and MD solution (1:1) (v/v)	Vacuum freeze drying-pre-frozen (-60°C for 6 hr) and vacuum freeze drying for 48 hr Ultrasound homogenizer-(power 200 W for 5 min)	EE (79.88%), AA (83.82% to 95.32%)	(Li et al., 2022)
Double drum drying	Red cabbage waste	MD20: GA (20:80)	Temperature (140°C), 5 rpm and 0.3 mm gap	EE (98.85%), anthocyanin (42.26 mg cyanidin-3-glucoside equivalent /100 g DW)	(Sakulnarmrat et al., 2021)

Technique	Waste material	Encapsu-lating agents	Conditions	Yield	References
Uniaxial electro-spinning	Spinach waste	(40% zein (6 g) 80% ethanol (15 mL)) (w/v) and (Traditional extraction of chlorophyll (TEC) (5mL)) (v/v) in a 3:1 ratio.	Flow rate (0.8 mL/h), applied voltage (17 kV), Distance to the collector plate (15 cm)	EE (57.13 to 82.95%)	(Isik et al., 2024)
		(40% zein (6 g) 80% ethanol (15 mL)) (w/v) and (Microwave extraction of chlorophyll) (MEC) (5 mL) in a 3:1 ratio.			
Coaxial electrospinning	Spinach waste	(TEC) and B- (30% zein (6 g) 80% ethanol (20 mL))	Flow rate (A: 0.3 mL/h B: 0.6 mL/h), applied voltage (17 kV), Distance to the collector plate (13 cm)	EE (49.53-126.64%)	(Isik et al., 2024)
		(MEC) and B- (30% zein (6 g) 80% ethanol (20 mL))			

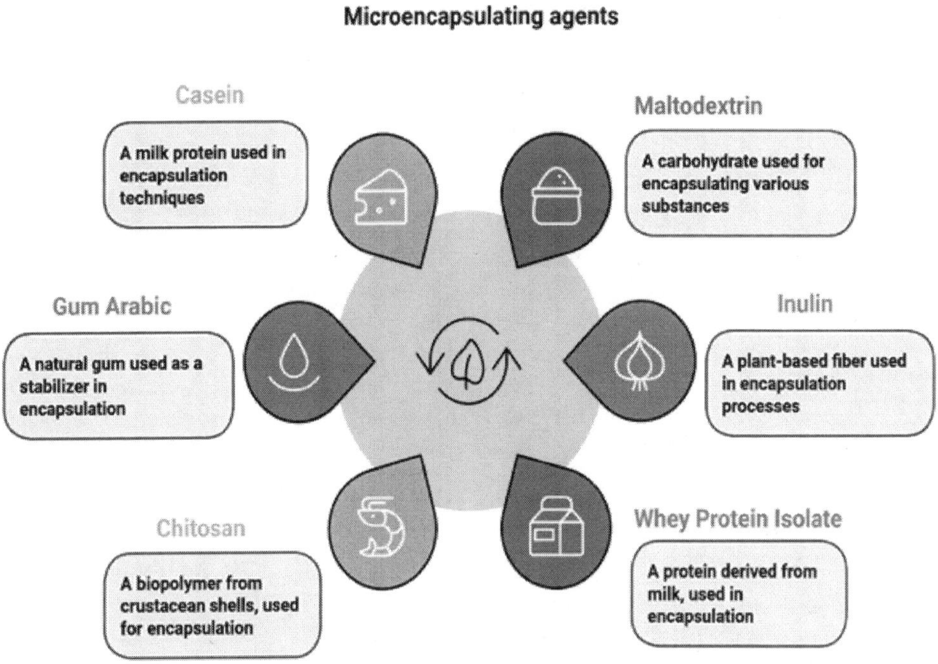

Fig. 1 Various microencapsulating agents used for spray drying and freeze drying

Valorization of Fruit and Vegetable Waste: Bioactive compounds its Applications in Functional Foods and Sustainable Innovations

Hariadi *et al.* (2023) utilized the dragon fruit peel waste for making natural dye. Dragon fruit skin contains anthocyanin pigment, which imparts a red color to dragon fruit peel. The method of encapsulation was done to preserve the natural pigment. The method included preparing the dragon fruit peel powder and encapsulating it with MD. The method used a completely randomized design of 10%, 20%, 30%, and 40% concentrations of MD and analyzed the total anthocyanin, water content, color saturation, and yield. The analysis parameter was higher at 30% concentration. The research emphasizes encapsulation's potential as a useful method for effective pigment preservation.

Naknaen et al. (2016) utilized watermelon rind powder (WRP) for developing cookies as FFs. The watermelon rind is rich in dietary fiber and BC. The study was done on the quality of cookies by replacing the wheat flour with WRP and high-maize starch (HMS). By using 20% of WRP and 30% of HMS concentration improves antioxidant property and increases the dietary fiber content; the same brings down the glycemic index to low in HMS and medium in WRP and produces a functional cookie.

Ahmed *et al.* (2021) enhanced the nutritional value of cakes by incorporating banana peel powder, a rich source of dietary fibers, minerals, and BC. This addition not only boosted the nutrient content but also introduced significant amounts of natural antioxidants like phenols, flavonoids, and carotenoids, which are beneficial for human health. Moreover, these antioxidants helped to prevent fat rancidity and possess antimicrobial properties, effectively extending the product's shelf life.

Nimbalkar *et al.* (2018) utilized the pea pod as a substrate for biobutanol production. The study states that pea is a good source of hemicellulose (21.1%) and cellulose (32.08%), which is categorized as a good source of carbon for biobutanol production. The pea pods were compositionally analyzed. The analysis includes fermentation, saccharification, drying study, and detoxification. Resulted in 4.25–5.94 g/L of total solvent with around 50% of sugar solution by the acetone-butanol-ethanol (ABE) fermentation of pea pods, which are detoxified. Overall, the ABE production was done by pea pod waste utilization.

Samotyja (2019) used the potato peel waste in the production of natural antioxidants. The solvent used for the extaction of phenolic compounds from potato peel is 96% and 80% of ethanol and water by use of model and antioxidant assays that consist of food lipids. Extract of 96% extracted highest phenolic compounds also showed the highest antiradical, iron-chelating, and reducing potential. Potato peel-extracted antioxidants showed higher efficiency than butylated hydroxyanisole (BHA) and butylated hydroxytoluene (BHT). The study states the utilization of potato peel for valorization and to increase the shelf stability of the product.

Kamath and R. (2018) investigated the potential of custard apple peel powder as an effective adsorbent for removing malachite green from aqueous solutions. By varying experimental conditions, they determined that optimal dye removal occurred at a slightly acidic pH of 6. The Highest removal efficiency was attained at an adsorbent dosage of 0.3 mg/50 mL and a contact time of 75 minutes, which was 95.3%. These results make the utilization of custard apple peel powder as a promising sustainable and low-cost alternative to mitigate malachite green-related water pollution effective.

Younis and Ahmad ((2015)) investigated to find the use of apple pomace powder as a functional ingredient in buffalo meat sausages. The proximate composition results for the pomace shown that, it contained 4.50% protein and 62.67% dietary fiber. More ever, its water- and oil-holding capacities as well as antimicrobial activity against *L. monocytogenes*, *P. aeruginosa*, and *S. aureus* were studied. Addition of apple pomace powder leads to increased dietary fiber content of the sausages, which favored higher stability of the emulsions and yields at cooking of 74.70% and 94.46%, respectively.

Bedrní ek *et al.* (2019) studied the onion waste as a natural antioxidant for the meat products. The antioxidant activity of cooked pork patties' on onion skin water extracts (OSWEs) was studied. The OSWE was added in the pork patties which enhanced the antioxidant property and which also helps in lowering the lipid oxidation, which was measured through thiobarbituric acid reactive substances. The two primary phenolic chemicals found in OSWEs were monoglucoside and quercetin. These findings indicates that onion skins can be a valuable source of natural antioxidants for their use in the food industry.

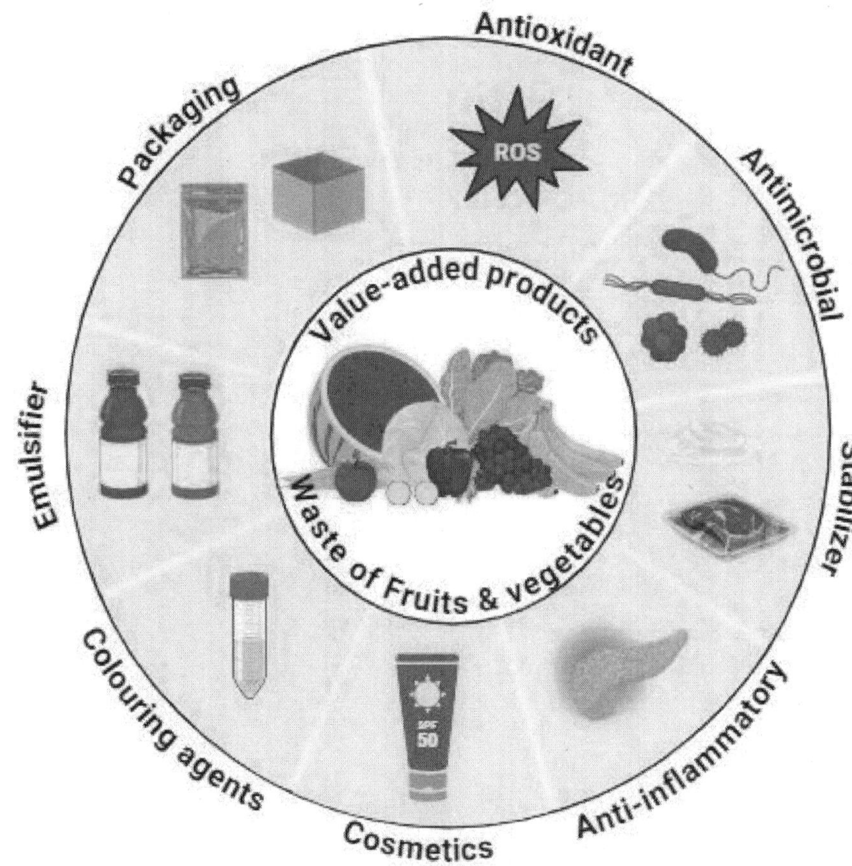

Figure 2. Versatile Applications of Bioactive Compounds Extracted from fruits and vegetable waste

Challenges and Future Prospects in the Valorization of Fruit and Vegetable Wastes by Microencapsulation

FVW from the food and agriculture sector can specifically lead to environmental damage because of the high microbiological load and moisture content. BC found in FVW degrade rapidly and even a minor change in extraction technique. Hence, it's important to stabilize bioactive components both before and after extraction and choose an optimized extraction technique for optimal quality (Nirmal et al., 2023). "The increasing requirement for natural food ingredients has led to increasing interest in extracting valuable compounds from vegetable and fruit byproducts. However, scaling up these processes from laboratory settings to industrial production is creating more challenges" (Taghian Dinani & van der Goot, 2023). Agri-food wastes offer a valuable opportunity for valorization, transforming them into high-value products like food additives, medicines, or other environmentally beneficial uses. These wastes are highly biodegradable and rich in bioactive chemicals, including antibiotics and pigments. The fruits and vegetable waste contains valuable phytochemicals like phenolics, carotenoids, and tocopherols that are widely used for nutritional, medicinal, and antioxidant purposes; Hence, this is one excellent component of ingredients to be used in the manufacture of FFs as well as pharmaceutical preparations for the chronic and acute illnesses. Therefore, the biomedical, cosmetic, and food sectors have a bright outlook in utilizing agri-food wastes (Osorio et al., 2021; Berenguer et al., 2022). One of the promising methods for valorizing fruit and vegetable waste is through the increasing stability and bioavailability of such bioactive compounds by using microencapsulation. The biggest challenges in this approach to comprise the use of cheaper and more biodegradable encapsulating materials that can cover sensitive compounds during processing and storage. The efficiency of encapsulation, release profiles, and stability depend on the wall material, which may comprise biopolymers. Scaling up from a lab to industrial scale when it comes to microencapsulation create a question about reproducibility and consistency in addition to high cost. The variability in waste composition and moisture content also affects the process. Future prospects for microencapsulation in waste valorization include advanced techniques like electrospinning and coacervation, as well as innovations in nanotechnology and food-grade materials. Combining microencapsulation with bioconversion and enzymatic processes could create new value-added products from fruit and vegetable waste.

CHAPTER 6

ECONOMIC ASPECTS OF WASTE VALORIZATION

The Economics of Agricultural Waste

Cost-Benefit Analysis of Waste Management vs. Valorization

Agricultural waste represents both a challenge and an opportunity for modern economies. Traditional waste management practices, such as landfilling and open burning, incur direct and indirect costs including transportation, land use, and environmental degradation (Smith et al., 2018). In contrast, waste valorization—the process of converting waste into valuable products—offers an economic advantage by creating revenue streams and reducing environmental costs (Prashant et al., 2020). Studies indicate that the net economic benefit of valorization can outweigh the costs of investment and operation, provided the processes are optimized (Jiang et al., 2021).

For example, converting rice husks into biochar not only mitigates the disposal cost but also generates a high-value soil amendment product. A comparative study in India revealed that valorization yielded a benefit-to-cost ratio of 2.5 compared to 0.7 for traditional disposal methods (Kumar et al., 2019). These findings emphasize the need for integrating economic analyses into waste management planning.

Economic Opportunities in Waste Valorization Industries

The global market for waste valorization industries is expanding, driven by advancements in technology and policy incentives. Waste-to-energy plants, biofertilizer production units, and bioplastics manufacturing facilities are emerging as key contributors to local and national economies (Huang et al., 2020). In the European Union alone, the bioeconomy's contribution to GDP is estimated at over 700 billion annually (Eurostat, 2021).

Small and medium enterprises (SMEs) also benefit from the growing demand for sustainable products. In Kenya, for instance, SMEs engaged in producing bio-briquettes from agricultural waste reported a 40% increase in profitability within five years of operation (Mwangi et al., 2022). These trends highlight the diverse economic opportunities inherent in waste valorization.

Market Potential for Valorized Products

Domestic and International Markets for Bio-Products

The market for valorized products is multifaceted, encompassing domestic and international segments. Domestically, demand for bio-based fertilizers, bioplastics, and biofuels is rising, driven by consumer awareness and regulatory frameworks promoting sustainability (Fernández et al., 2019). Internationally, valorized products cater to niche markets such as organic agriculture and eco-friendly packaging. The export of bioplastics from Southeast Asia to Europe grew by 30% annually between 2015 and 2020 (Lee et al., 2021).

One notable example is the export of biochar from Vietnam, where agricultural residues like coconut shells are processed into high-quality products for international markets. The success of such initiatives underscores the potential for developing countries to tap into global value chains (Nguyen et al., 2020).

Case Studies of Successful Valorization Businesses

Case studies offer valuable insights into the practicalities of waste valorization. A prominent example is the "AgroCycle" initiative in Europe, which has successfully commercialized products such as bioplastics and organic fertilizers derived from agricultural waste (Garcia et al., 2018). Similarly, "EcoGreen Farms" in India leverages cow dung and crop residues to produce biogas, generating annual revenues of $2 million (Sharma et al., 2021).

These examples illustrate the importance of innovation, market understanding, and supportive policies in scaling waste valorization businesses. They also demonstrate how collaborations between academia, industry, and government can enhance the economic viability of such initiatives.

Financing and Investment Opportunities

Funding Options, Subsidies, and Incentives

Securing funding is critical for waste valorization projects. Multiple sources, including government grants, private equity, and international development funds, are available to support these initiatives. For instance, the European Union's Horizon 2020 program allocated €3 billion to bioeconomy projects between 2014 and 2020 (European Commission, 2021).

Subsidies and tax incentives also play a crucial role. In Brazil, sugarcane ethanol production benefits from reduced taxes and direct subsidies, making it a profitable venture (Silva et al., 2020). Similarly, India's National Biogas and Manure Management Program provides capital subsidies to small-scale biogas plants (Government of India, 2021).

Public-Private Partnerships and Their Role in Waste Valorization

Public-private partnerships (PPPs) have emerged as an effective mechanism for financing and implementing waste valorization projects. By leveraging the expertise and resources of both sectors, PPPs address challenges such as high initial capital costs and technological complexity (Reddy et al., 2020). For example, the "Green Waste Energy" initiative in South Africa is a PPP that converts municipal waste into electricity, serving over 100,000 households (Johnson et al., 2019).

The success of PPPs often depends on transparent governance, equitable risk-sharing, and long-term policy support. These partnerships not only enhance financial viability but also contribute to broader socio-economic goals, including job creation and energy security.

Challenges and Risks

Market Entry Barriers, Policy Issues, and Economic Risks

Despite its potential, waste valorization faces significant barriers. Market entry challenges include high capital requirements, lack of technical expertise, and competition from conventional products (Mehta et al., 2019). Policy inconsistencies, such as abrupt changes in subsidies or regulatory standards, further exacerbate these challenges.

Economic risks, including fluctuating prices of raw materials and finished products, also pose threats to the sustainability of valorization projects. For instance, the volatility of crude oil prices impacts the competitiveness of biofuels, as seen during the COVID-19 pandemic (Singh et al., 2021).

Strategies for Overcoming These Challenges

Strategies to mitigate these risks include fostering innovation through research and development (R&D), creating stable policy environments, and building robust supply chains. Governments can play a pivotal role by offering predictable incentives and facilitating access to markets through trade agreements (Zhang et al., 2021). Additionally, capacity-building programs can address skill gaps and promote entrepreneurship in the sector.

Collaborative platforms that bring together stakeholders across the value chain are also essential. For example, the "Circular Bioeconomy Alliance" fosters global collaboration to address systemic challenges in waste valorization (Brown et al., 2020).

CHAPTER 7

ENVIRONMENTAL IMPACT OF AGRICULTURAL WASTE VALORIZATION

Waste valorization is the process which is used for turning waste materials into useful products, known as waste valorization, it improves both environmental and economic sustainability. This technique is phenomenal to the circular economy, which helps to reduce waste accumulation and also increases resource efficiency by transforming trash from one process as a resource for another process *(Durandeau et al., 2009)*. By utilizing organic, agricultural, and industrial waste into value-added products like biofuels, bioplastics, fertilisers, and biocatalysts, waste valorization is the strategy which reduces the amount of waste that is dumped in landfills, mitigate greenhouse gas emissions, conserve resources, and helps to improve soil health *(Angili et al., 2022)*. Waste generation forms major environmental risks, including air pollution, water contamination, and soil deterioration, all of which it creates a disbalance between ecosystems and compromises public health. It monstly contains organic residues that can emit hazardous compounds into soil and water and generate by creating greenhouse gases, unmanaged and untreated agricultural waste in particular manner which can implement a serious environmental danger. By valorizing organic and agricultural waste into renewable, sustainable energy sources and products which will improve soil health. Waste valorization helps to mitigate these most concerning factors *(Falade, 2021)(Tropea, 2022)* .

Waste valorization, it is a fundamental element of the circular economy, which aims to convert waste into valuable and sustainable resources, which also helps in lowering landfill waste and boosts resource efficiency by improving economic and environmental sustainability *(Sommer et al., 2016)*. According to Sharma et al. (2016), repurposing food, industrial, and agricultural waste into biofuels, bioplastics, fertilisers, and biocatalysts greatly helps to lower the greenhouse gas emissions, also preserves natural resources, and can be utilised to enhance soil health. Ecosystems and public health are singnificantly affected by the environmental risks related with unprocessed garbage, which include soil erosion, water pollution, and air contamination (Stavi & Lal, 2012). Utreated and unmanaged agricultural waste enhances a serious concern to the environment because it can contaminate soil and water and also generate hazardous greenhouse gases. By valorizing organic waste into renewable energy and soil-enriching products, this technique reduces the hazards and also helps to advance the Sustainable Development Goals (SDGs) of the United Nations, such as SDG 12 (Responsible Consumption and Production) and SDG 13 (Climate Action) (Guo, 2020). By utilizing valuable materials in production cycles, valorization acts as a sustainable strategy which lessens the environmental impact of waste and motivates to apply resilient agricultural practices (Angili et al., 2022).

The main goal of using this technique is to reduce waste's negative environmental effects by altering it into sustainable resources. In particular, it can be used to promote carbon sequestration, improving the soil health, mitigation of soil carbon depletion, and supports sustainable production methods with significant effect on lowering greenhouse gas emissions. By incorporating waste-to-useful resource strategies that used to reduce environmental impact and applies to increase resource efficiency, these targets are in line to follow the SDGs approved by UN, mostly SDGs 12 and 13 *(Tropea, 2022).*

Reducing Environmental Footprint

Reducing greenhouse gas emissions and increasing soil health through creative and value added processes that modify waste into biofuels, biofertilizers, and soil amendments, waste valorization can also be used for lowering environmental footprints and promotes sustainability in most of the sectors, mainly in the energy and agriculture sectors (*Kodba, Pukšec & Dui´c.,2023*) (Angili et al., 2022).

Mitigating Greenhouse Gas Emissions

Food and agricultural generated wastes are the main source of greenhouse gas emissions, mostly Nitrous oxide (N2O) and Methane (CH□), which have a greater potential to cause global warming than carbon dioxide (CO_2). While nitrous oxide is majorly emitted from excessive use of nitrogen fertiliser used in agriculture (falade,2021), Methane emissions in environment are produced by the anaerobic breakdown of organic waste generated in landfills *(Tropea, 2022).* The formation of biogas by anaerobic digestion (AD) is one significant valorization application to reduce greenhouse gas emissions. Food waste and agricultural waste are processed by AD into biogas, it is a renewable energy source that can be utilised for the reduction of carbon emissions by replacing the place of fossil fuels. According to research, processing biogas from agricultural and food waste and manure can mainly decrease greenhouse gas emissions by up to 80%, helping the EU fulfil its emission reduction targets (Angili et al., 2022) *(Kodba, Pukšec & Dui´c.,2023) (Tropea, 2022)*. By processing nutrient-rich digestate as a value added by-product, which can be reutilised to the soil to increase it's fertility and decrease N_2O emissions, biogas production not only focuses on lowering methane emissions but also lowering the demand for chemical fertilisers (Bhatnagar, 2015).

When formulating the environmental advantages of waste-to-energy technologies, life cycle analysis, or LCA, is very much essential. LCA provides a descriptive understanding of sustainability by assessing the inputs, outputs, and environmental effects at each and every step of the analysis and valuation process. Enzyme loading and solvent consumption in fat extraction, for example, were detected as environmental hotspots in a laboratory-scale life cycle assessment (LCA) of food waste valorization into bioethanol, biomethane, and bio-oil. This shows the optimisation at these phase for enhanced sustainability (Angili et al., 2022). LCA of food waste bioconversion into biofuels showed major greenhouse gas reductions, as reported by (*Angili et al., 2022*), accepting the potential of biorefinery technologies to generate most of the environmental benefits when staged up (*Angili et al. 2022*).

The application of policy-driven objectives are most crucial to for reducing emissions. The effects of excess nitrogen fertiliser on lowering GHG emissions can be identified using economic and sustainable models, likewise those that combine biophysical and economic simulations. Research performed in northern France using such economic models (AROPAj) and the biophysical models (CERES-EGC) has showed that certain regulations on the usage of nitrogen fertiliser mainly lower N2O emissions. According to (*Durandeau at al., 2009*), this method makes it possible to confirm the potential for emissions reduction at the farm level. Cost-effective reductions in greenhouse gas emissions result from optimal fertiliser application (Kodba, Pukšec & Dui´c.,2023).

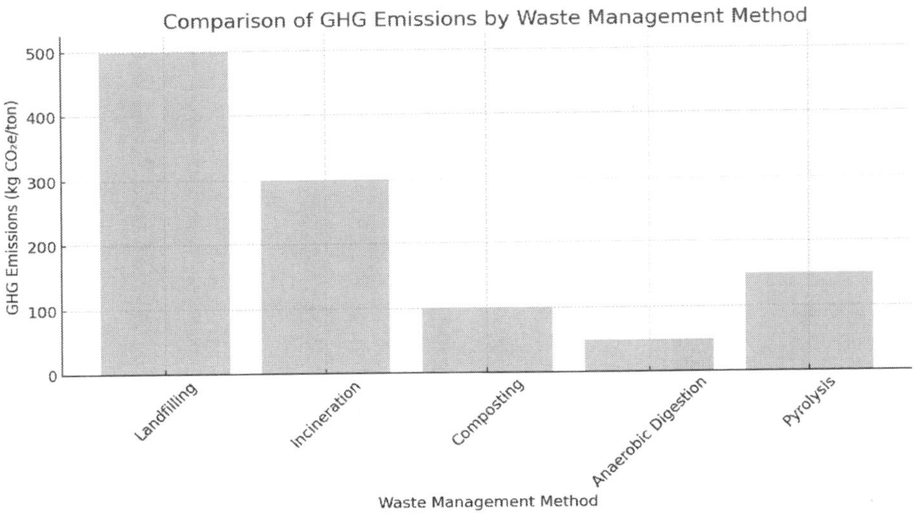

Fig 1.1 Comparison Of GHG Emissions By Waste Management Method

From Fig 1.1 The environmental effects of conventional and traditional waste disposal methods like landfilling and incineration in comparison to the newly valorization methods like composting, anaerobic digestion, and pyrolysis are expressed in the bar graph, which compares greenhouse gas (GHG) emissions from different waste management processes. Methane, which is a powerful greenhouse gas with a global warming potential far greater than CO_2 and other GHG gasses, which is released in tremendous quantities during the anaerobic decomposition of organic material and waste in landfills, which have the largest GHG emissions at about 500 kg per tonne of waste (IPCC, 2007). The exhaustion and combustion of fossil-based materials in waste like food and agricultural waste results in significant emissions of CO_2 and other pollutants and GHG gases, with an average of 300 kg CO_2 tonne, even if the incineration is more efficient than landfilling with respect to lowering the volume of waste (Gentil et al., 2010).

On the other hand, composting uses aerobic processes which reduce methane formation, it only generate about 100 kg CO_2 per tonne, while emissions might change based on waste content and management techniques used (Morris, 2005). Amongst all these processes, anaerobic digestion has the lowest emissions (50 kg CO_2 per tonne), which makes it particularly environmentally friendly. This procedure lowers the complete environmental impact by capturing methane for use as a renewable energy source (IPCC, 2007; Gentil et al., 2010). The thermal decomposition process known as pyrolysis, which yields biochar, bio-oil, and syngas,which has modest greenhouse gas emissions (150 kg CO_2 per tonne) as per the other techniques and may help in carbon sequestration if the biochar is incorporated into soil, it leads a way in long-term carbon storage (IPCC, 2007).

So we can easily conclude that the anaerobic digestion and composting which generally have lower greenhouse gas emissions than most of the traditional and conventional disposal methods, this comparison mainly focused on the environmental benefits of waste valorization approaches. Hence the most sustainable technique is anaerobic digestion, pyrolysis and composting since it can generate renewable energy, and also shows how waste valorization can enhance resource recovery and reduce environmental impacts (Morris, 2005; Gentil et al., 2010).

Soil Health Improvement and Carbon Sequestration

This technique requires enhancing carbon sequestration and improving soil health for environmental sustainability. While soil carbon sequestration enhances the combat climate change by using methods like absorbing and storing atmospheric carbon, soil health is essential and crucial for agricultural output and environmental resilience.

Biochar—a carbon-rich substance—improves soil structure, water retention, and nutrient availability. In addition it also increases soil fertility, biochar acts as a long-term carbon sink which significantly reduces CO2 emissions also used as a soil amendment. According to studies, using biochar can mostly improve soil characteristics and increase it's soil carbon storage, especially in nutrient-poor soils (falade,2021) *(Tropea, 2022)*. Another technique for generating value is composting, which coverts organic waste into soil beneficial nutrients that are high in nutrients. It also increases soil water retention and microbial activity by increasing soil organic matter, both of which are advantageous in respective areas. Compost and biochar can also increase soil resistance to environmental stress, it lowers reliance on chemical fertilisers and also promotes long-term agricultural sustainability, according to (*Falade , 2021*).

Soil health and nitrogen cycling are majorly influenced by the microbial activity in soils. Compost and biochar are examples of organic additions which boost microbial variety and activity, it also promote plant nutrient availability and improves the soil's capacity to store carbon. In the breakdown of organic matter, it promotes nutrient cycling, and enhances soil structure. Enzymes like ligninolytic and cellulolytic enzymes that are derived from the valorization of agricultural waste. *(Tropea, 2022)* explains how microbial enzymes from waste valorization can help to lower soil pollution levels, also promotes improved crop yield and healthier ecosystems *(Tropea, 2022)*.

Valorization of food and agricultural waste as a resource, soil has many benefits for the environment and the economy. Production processing cost are to be reduced and nitrogen runoff pollution is also reduced by utilizing organic amendments made from valorised waste inspite of the synthetic fertilisers. In addition soils that have been improved with organic carbon are more resistant to drought and erosion, which helps to promote sustainable farming methods. Economic modelling explains that waste valorization also improves soil health, by offering long-term cost savings, reducing dependency on chemical inputs, and promoting ecological resilience (Durandeau et al., 2010).

It is a sustainable technique which used for turning waste into useful resources, encouraging the use of renewable energy, by enhancing soil health, and by reducing greenhouse gas emissions in waste valorization. These techniques provide a cleaner environment and a more improved agricultural system while supporting the global sustainability goals.

Waste Valorization and Circular Economy

A key component of waste valorization is circular economy, majorly in agriculture, where organic waste products are being recycled into sustainable and value added goods. The circular economy framework is a strong technique which focuses on minimising waste, lowering dependency on virgin resources, and keeps materials in the production line. By producing bioenergy, biobased products, and soil amendments this technique support closed-loop nutrient and energy cycles, the value-addition of organic waste and agricultural residues promotes sustainable farming *(shah, Venkatramanan & prasad.,2021)*.

Integration into Circular Agricultural Practices

Conversion of trash from one phase of processing into a resource for another processing phase, a circular agricultural model can reduces its harmful effects on the environment and promoting sustainable land use. To improve soil health and nutrient content, it is essential and important to recycle crop leftovers, animal manure, and the organic waste promoting into agricultural applications as biofertilizers and soil supplements.

Nutrient-dense biofertilizers which contain essential elements like nitrogen, phosphorus, and potassium are formed from waste valorization processes like anaerobic digestion and composting. These biofertilizers enhances the soil fertility and structure and lowering the demand for chemical inputs *(shah, Venkatramanan & prasad.,2021).(Kodba, Pukšec & Dui´c.,2023)* Energies. Various research contains that biofertilizers made from agricultural and food waste improves the nutrient cycling in soil systems with boosting crop yields, both the techniques are essential for preserving and enhancing soil fertility with respect to long term vision (Duan, 2021). By processing trash into a renewable supply of plant nutrients, by lowering dependency on synthetic and non organic fertilisers, and by lowering environmental pollution linked to fertiliser use in excess amount, by using biofertilizers constituently with the concepts of the circular economy (Aslam & Nazir.,2024). Using biofertilizers it enhances soil biological activity, which is most important for microbial biodiversity and nutrient processing, resulting in effective agricultural ecosystems *(shah, Venkatramanan & prasad.,2021).*

Within the processing area of the circular economy, the production of biogas from agricultural waste, such as crop leftovers and manure, provides a feasible renewable energy source. Anaerobic digestion transforms organic waste into biogas, which can be used as fuel for cars, and for heating, to extract and generate power. Digestate, it is a by-product of this process, which is a great biofertilizer that completes the nutrient cycle by fortifying the soil with important nutrients *(Kodba, Pukšec & Dui´c.,2023)*. Farms can be managed by utilising organic waste sustainably and by lowering their carbon footprint by accepting to biogas from fossil fuels. It is a key element method of agricultural waste valorization, biogas is recognised by the European Union's Renewable Energy Directive as a substantial contributor to greenhouse gas reduction when produced sustainably *(Kodba, Pukšec & Dui´c.,2023)* (rajmohan, Ramya & sunita.,2019).

Table 1.1 Agricultural Waste Valorization and Benefits

Agricultural Waste	Valorized Products	Benefits	References
Rice Husk	Biochar, Silica, Rice Husk Ash (RHA)	Soil amendment, carbon sequestration, renewable energy	IPCC, 2007
Sugarcane Bagasse	Bioethanol, Biogas, Charcoal	Renewable energy, reduction of fossil fuel use, soil improvement	Gentil et al., 2010

Agricultural Waste	Valorized Products	Benefits	References
Fruit and Vegetable Peels	Organic fertilizers, Animal feed	Enhanced soil fertility, waste reduction, improved livestock nutrition	Morris, 2005
Corn Stover	Biofuels (e.g., ethanol), Bioplastics	Renewable fuel source, reduced plastic pollution	Li et al., 2009
Wheat Straw	Cellulose, Paper, Bioethanol	Sustainable materials, reduced deforestation, renewable energy	Zhang et al., 2011
Coconut Shell	Activated Carbon, Charcoal	Water purification, renewable energy	Kumar et al., 2010
Coffee Grounds	Biodiesel, Bioplastics	Reduction of plastic pollution, renewable energy	Kondamudi et al., 2008
Olive Pomace	Polyphenols, Biofuels	Antioxidant production, renewable energy	Roig et al., 2006

From Waste to Resource

Waste valorization is the the foundation of a linear to a circular economic model is the conversion of agricultural waste into useful and value added products. In this technique transition is biorefineries, by which it is possible to separate bio-based chemicals, biofuels, and bioplastics from agricultural waste. This strategy supports and encourages both environmental sustainability and economic growth by enforcing the development of sustainable materials to replace petrochemical-based products (*shah, Venkatramanan & prasad.,2021*) (121).

Cellulose, hemicellulose, and lignin are found in agricultural wastes like lignocellulosic residues (straw, corn stover), and these substances play important role as a raw materials for generating bioproducts like organic acids, bioplastics, and bioethanol. These materials can be converted and reconsider into sustainable alternative for plastics and other synthetic materials by using bioconversion techniques like fermentation and enzymatic hydrolysis (*shah, Venkatramanan & prasad.,2021*). Biorefineries make it sure to process a variety of feedstocks effectively, and producing renewable energy and high-value added byproducts. For example, lignin which can be used in bio-based adhesives and composites, while cellulose which has been extracted and isolated from crop leftovers which can be further used to create biodegradable packaging. By assuring that different agricultural wastes are fully utilised and converted as a value added product within the circular bioeconomy framework this multi-product method enhances the economic feasibility of biorefineries. (*shah, Venkatramanan & prasad.,2021*).

Compost and biochar, are two different organic techniques made from agricultural waste, are essential and effective for managing soil health. Pyrolysis-produced biochar improves soil water retention and nutrient retention which helps for providing long-term carbon sequestration benefits. In addition to offer a nutrient-rich soil amendment, composting organic waste also promotes in waste management and lowers greenhouse gas emissions from decomposition (*shah, Venkatramanan & prasad.,2021*) (Das, ghosh & avasthe.,2020). Because of the benefits, in valorization technique biochar mainly useful in damaged and non fertile soils because it enhances microbial activity and reduces nutrient loss from the soil. On the other hand, composting which encourages nitrogen cycling and enhances the soil organic matter, both of which are essential and easily applicable for sustainable agriculture practices. Utilising agricultural waste to improve and enhance the soil productivity, these techniques promote a sustainable and regenerative agricultural system, which is relevant and sustainable with the concepts of the circular economy (Aslam & Nazir.,2024).

Addressing Pollution and Resource Depletion

Valorization of agricultural and food waste into useful resources and reducing it's negative, hazardous effects on the environment, effective waste valorization minimizes the problems of various pollution and resource depletion. Sustainable waste management techniques are crucial and essential for reducing pollution and preserving vital resources such as soil nutrients, water, and air quality, particularly in agricultural systems. Anaerobic digestion, the use of biochar, and biorefining are the prime techniques of waste valorization methods that help to reduce pollution and promote the necessary resource recovery (Barros et al.,2020) (Bharagava & saxena.,2020).

Prevention of Water, Soil, and Air Pollution

Agricultural runoff without treatment leads to severe effects on water pollution, especially regarding fertilizer leaching and the subsequent poisoning of aquatic systems through nitrogen and phosphorus. Such an issue would be diminished by using bio-based solutions like valorized biofertilizers. Biochar and many organic amendments that enhance water retention and prevent nutrient leaching can protect water resources against alarming effects of agricultural pollutants (Bharagava et al. 2018). For example, digestates produced during anaerobic digestion can be used in fertilizing agriculture to restore the quality of nutrients within soil and reduce the use of synthetic fertilizers towards reduction of nutrient runoff. When applied with great care, such digestates have the ability to improve soil health and water quality as slow-release fertilizers with the reduction of surpluses nutrient loads from contaminating the adjacent water bodies (*shah, Venkatramanan & prasad.,2021*) (Adegbeye et al. 2019). Preparing biochar from agricultural wastes is a useful tool in soil remediation and pollution reduction through the process of pyrolysis. Its higher carbon content and pore nature constitute appropriate medium adsorption for harmful contaminants, such as pesticides and heavy metals, thus preventing their percolation into groundwater and soil. Biochar enhances soil fertility and health by improving microbial activity, pH, and water-holding capacity for consistent use. According to research, biochar has the potential to immobilise pollutants in soil, which makes it a most important method of controlling agricultural pollution (*shah, Venkatramanan & prasad.,2021*) (Bharagava & saxena.,2020).

Anaerobic digestion (AD) agricultural and food waste produces biogas, a source of renewable energy which can replace fossil fuels. This approach also avoids methane emissions from organic wastes that would otherwise decompose in open environments. Moreover, apart from curbing greenhouse gas emissions, AD technology mitigates air pollution associated with traditional sources of energy through the capture of methane in the biogas production process (*shah, Venkatramanan & prasad.,2021*). Moreover, the use of biogas for energy development cuts the carbon footprint of farms and fosters a closed-loop system in which waste products are applied optimally.

Sustainable Resource Use and Conservation

Many problems associated with the present agriculture are resource depletion by overuse synthetic fertilizers, water, and fossil fuels that have catalyzed unsustainable practices. Waste valorization operates on a basis of circular economy premise where agricultural residues are used efficiently, therefore conserving natural resources. For example, biorefineries convert the agricultural residues into biofuels, biochemicals, and bio-based materials, which replace the traditional conventional petroleum-based products and their dependence on non-renewable resources (Lytras et al. 2020) (Barros et al.2020).The concept of biorefinery also encourages the production of several high-value products from one feedstock, thereby increasing resource efficiency. Lignocellulosic biomass, which encompasses crop residues such as corn stover and rice husks, can be converted into biofuels, biodegradable plastics, among many other green materials, significantly reducing the requirement for new raw materials, and supporting the principles of a circular economy (*shah, Venkatramanan & prasad.,2021*) (Barros et al.2020).

Valorization of waste through organic amendments such as compost and bio-fertilizers helps in maintaining the fertility of soil and reduced dependence on chemical fertilizers. Organic amendments result in sustainable nutrient cycling as nutrients are released slowly, thus improving the structure and health of the soil in the long run. Digestates from anaerobic digestion and biochar supply essential amounts of nutrients to the soil and act as excellent alternatives to synthetic fertilizers for enhancing long-term soil productivity and conservation of soil resources (Adegbeye et al. 2019).

Environmental Challenges

Valorization can sometime create various additional problems with emissions, pollution, and energy consumption even though its final and ultimate goal is to reduce environmental harm. It act as a balanced The problems of converting various waste types—particularly culinary, industrial, and agricultural waste—into goods with value added to give rise to the environmental issues related to the waste valorization technique which optimizes positive effects with reducing negative ones (*shah, Venkatramanan & prasad.,2021*) (Sarker et al.,2023).

Potential Negative Environmental Impacts

Anaerobic digestion and pyrolysis are two efficient waste-to-energy processes that can turn garbage into biogas or biochar, but they also use a lot of energy and can release emissions. For instance, methane and carbon dioxide are byproducts of anaerobic digestion. Inadequate containment measures can allow methane, a powerful greenhouse gas, to escape into the atmosphere. If renewable energy sources are not used, the total environmental benefits of pyrolysis and gasification—which need high temperatures—will be diminished because they also use a significant amount of energy, frequently from fossil fuels (Gil,2021) (Lee et al.,2019). According to life cycle analyses (LCAs), the process of turning food waste into biofuels imposes some environmental costs, particularly when hydrolysis and the use of chemical solvents are involved. The environmental impact of producing biodiesel is greatly increased by these procedures, underscoring the need for more effective and clean solutions. According to research, solvent-based extraction may release volatile organic compounds (VOCs) that have an impact on air quality, and some processing additives may produce secondary pollutants (*Angili et al., 2022*) (Arancon et al.,2013).

Remaining pollutants from the original waste material are frequently included in biofertilizers, which are byproducts of waste valorization. For instance, trace heavy metals or persistent organic pollutants (POPs) from industrial byproducts or municipal solid waste may be present in digestates from anaerobic digestion. These pollutants can build up in soils if they are not properly treated, which over time will deteriorate soil quality and damage crop health (Barros et al.,2020) (Arancon et al.,2013). Aquatic habitats are under risk due to nutrient leakage and eutrophication caused by runoff from these biofertilizers (Sarker et al.,2023). Composting and the use of biofertilizer are two waste valorization techniques that involve microbial processes that, if not controlled with strict biosecurity measures, could unintentionally release pathogens. Bacteria, fungi, and viruses may be present in food and agricultural wastes and may endure treatment procedures. Because insufficient treatment may reintroduce these bacteria into agricultural soils, pathogenic hazards are particularly significant for food waste valorization (*shah, Venkatramanan & prasad.,2021*).

Table 1.2 Environmental Impacts and Mitigation Strategies in Waste Management

Negative Environmental Impact	Mitigation Strategy	References
Greenhouse Gas Emissions	Implement waste valorization techniques like composting and anaerobic digestion to reduce methane emissions.	IPCC, 2014.
Water Pollution	Adopt wastewater treatment and phytoremediation techniques to remove contaminants before release into water bodies.	Wang, X., et al. 2019.
Soil Degradation	Use biochar and compost as soil amendments to improve soil health and reduce nutrient leaching.	Lehmann, J., et al. 2011.
Air Pollution from Incineration	Utilize advanced air filtration systems and reduce waste burning through recycling and valorization.	Nanda, S., et al. 2021.
Biodiversity Loss	Implement sustainable waste practices and reduce landfill use to protect ecosystems and habitats.	Pereira, H.M., et al. 2012.
Resource Depletion	Adopt circular economy practices, promoting recycling and reuse to conserve resources.	Geissdoerfer, M., et al. 2017.

Mitigation Strategies and Sustainable Practices

Implementing sustainable practices, technology advancements, and legislative assistance that collectively address emissions, contamination, and efficiency are necessary to mitigate the environmental difficulties connected with waste valorization. Environmental effects can be considerably decreased by streamlining the valorization process. Pre-treating feedstocks and carefully regulating pH and temperature during anaerobic digestion can increase methane production efficiency while lowering methane emissions. Closed-loop systems and sophisticated bioreactors are also efficient in reducing leaks and capturing emissions (Sarker et al.,2023) (Dhanya et al.,2020). By integrating renewable energy sources to power high-temperature processes, pyrolysis and gasification systems can lessen their reliance on fossil fuels (Barros et al.,2020). Furthermore, using biocatalysts and green solvents in the manufacturing of biofuel provides a more environmentally friendly option than using traditional chemicals. For instance, harsher chemical hydrolysis techniques are replaced by enzymatic hydrolysis, which lowers the production of dangerous byproducts. These developments reduce the environmental costs of waste valorization and are consistent with the principles of green chemistry (Dhanya et al.,2020).

Before being applied to soils, biofertilizers made from waste should be processed to lower the amount of heavy metals and other contaminants to avoid contamination. While soil testing and monitoring can track pollutant levels over time, standards for maximum contaminant levels in biofertilizers can help assure safe application (Arancon et al.,2013). Optimising the usage of biofertilizer by applying the proper proportions to reduce leaching and nutrient run-off *(shah, Venkatramanan & prasad.,2021)* is another benefit of precision agricultural techniques. One effective method for evaluating and reducing the environmental effects of waste valorization systems is life cycle analysis (LCA). Through the identification of high-impact phases in the value chain, life cycle assessment (LCA) can direct process enhancements that reduce waste and emissions. For example, a thorough life cycle assessment (LCA) of the manufacture of biofuel from food waste revealed that a high enzyme usage rate during hydrolysis is a major contributor to environmental impacts.

The total footprint could be decreased by lowering the dosage of enzymes or by using less harmful enzymes *(Angili et al., 2022)*. It's also crucial to continuously monitor the ecology around waste valorization facilities. Monitoring the quality of the soil, water, and air enables adaptive management techniques that stop long-term environmental deterioration by detecting and reducing pollutant levels in real-time (Arancon et al.,2013).

Strong legislative frameworks are essential for encouraging sustainable waste valuation. Regulatory initiatives that promote greener practices include criteria for biofertilizer pollutants, stringent emission rules, and incentives for employing renewable energy in waste processing. Policies under the Renewable Energy Directive in the European Union encourage the production of sustainable bioenergy and encourage adherence to strict objectives for reducing greenhouse gas emissions (Barros et al.,2020) *(shah, Venkatramanan & prasad.,2021)*. Innovation and adoption can also be encouraged by financial incentives like tax breaks for green waste valorization procedures and subsidies for environmentally friendly technologies. A further factor in accomplishing environmental objectives is funding research and development in sustainable waste valorization technology(Sarker et al.,2023).

Conclusion

Finally, in order to address the complex environmental issues related to agricultural waste, waste valorization is crucial. By reintegrating trash into production cycles, waste valorization—the process of turning agricultural leftovers and by-products into valuable resources—supports the circular economy by lowering greenhouse gas emissions, conserving natural resources, and reducing the amount of garbage that ends up in landfills. The United Nations Sustainable Development Goals, which place a strong emphasis on responsible production, consumption, and climate action, are among the major sustainability goals that this strategy supports.

Anaerobic digestion, composting, and the creation of biochar are examples of waste valorization processes that provide useful ways to turn trash into soil amendments, biofuels, and biofertilizers. These value-added products improve soil health, produce sustainable energy sources, and lessen reliance on synthetic fertilisers. Methane and carbon dioxide emissions are greatly decreased and renewable biogas is produced when waste materials are converted into energy through anaerobic digestion. In addition to lowering greenhouse gas emissions, this strategy helps the agriculture industry adopt sustainable energy sources, which lowers its carbon footprint. According to life cycle assessments (LCA), bioenergy made from trash can significantly lower greenhouse gas emissions, supporting international emissions reduction goals and advancing the low-carbon agricultural paradigm.

Trash valorization makes it possible to recycle and reuse agricultural byproducts within the framework of the circular economy, converting trash into a sustainable supply of materials, energy, and nutrients. Biorefineries, for example, play a key role in this transition by removing biochemicals, bio-based products, and biofuels from lignocellulosic wastes like maize stover and straw. By using every part of the feedstock, this technique reduces waste in addition to substituting bio-based products for petroleum-based ones. By ensuring that by-products, like digestates from anaerobic digestion, are returned to the soil as biofertilizers, biorefineries' closed-loop design completes the nutrient cycle and lessens the need for artificial fertilisers.

Waste valorization is a revolutionary method of handling agricultural waste that makes a substantial contribution to resource conservation, environmental preservation, and sustainable agriculture. trash valorization lowers agriculture's environmental impact, conserves resources, and promotes a resilient, circular agricultural model by turning trash into bio-based goods, renewable energy, and organic supplements. Despite ongoing obstacles, waste valorization will be able to realise its full potential with the help of technical advancement, governmental assistance, and sustainable practices, which will help create a more resilient, resource-efficient, and clean agricultural system. garbage valorization's economic and environmental advantages highlight how crucial it is to the shift to a sustainable future where garbage is valued as a resource rather than a burden.

CHAPTER 8

ROLE OF SOCIAL AND CULTURAL DIMENSIONS IN WASTE VALORIZATION

Engagement in Waste Valorization, Role of local communities in waste collection and processing

Social benefits: job creation, skill development, and empowerment

Before the study on the food waste management we should first learn about what is the food waste.

Different researcher or workers has adopted different criteria for defining the food waste. Whereas (Brian et al., 2013) defined food waste as "the food that is of acceptable quality and qualified for human consumption but is not consumed because it is squandered either before or after it is deteriorates."

The food waste has become the major global concern to the environment, economic as well as food security. When this food is wasted in large amount it leads to the emission of certain gases like methane, which is an extreme potent greenhouse gas is produced. The cost to reproduce the food also certainly increase the financial concern or burden in the food industry. In the context of the food security production of food waste leads to increase the country's dependency on the imported foods as its decrease amount of food available to be consumed (Sharma et al., 2021).

In addition, it was estimated that the total greenhouse gas (GHG) emission from the solid waste was approximately 5% of the overall GHG gas emissions into the atmosphere (Agrawal et al., 2021).

The food waste due to the processing also caused by losses in the nutritional value, calorific value, and the edibility of the crops caused by the extreme temperature range, high humidity or the action of unwanted microbes. Heat along with the high humidity creates a favorable grounds for the breeding of the pests which is the common cause of the food waste during storage (FAO 2012). A significant amount of the food waste are being generated at the retailer's level. The fishing business waste a lot of foods; in Europe between 40-60% of fish are wasted because of their inappropriate kind or shape and size (Stuart 2009).

Food waste can be utilized as a feed stock for the butanol fermentation because it contains carbohydrate in high amount (Smithers 2008; kosmala et al., 2011).

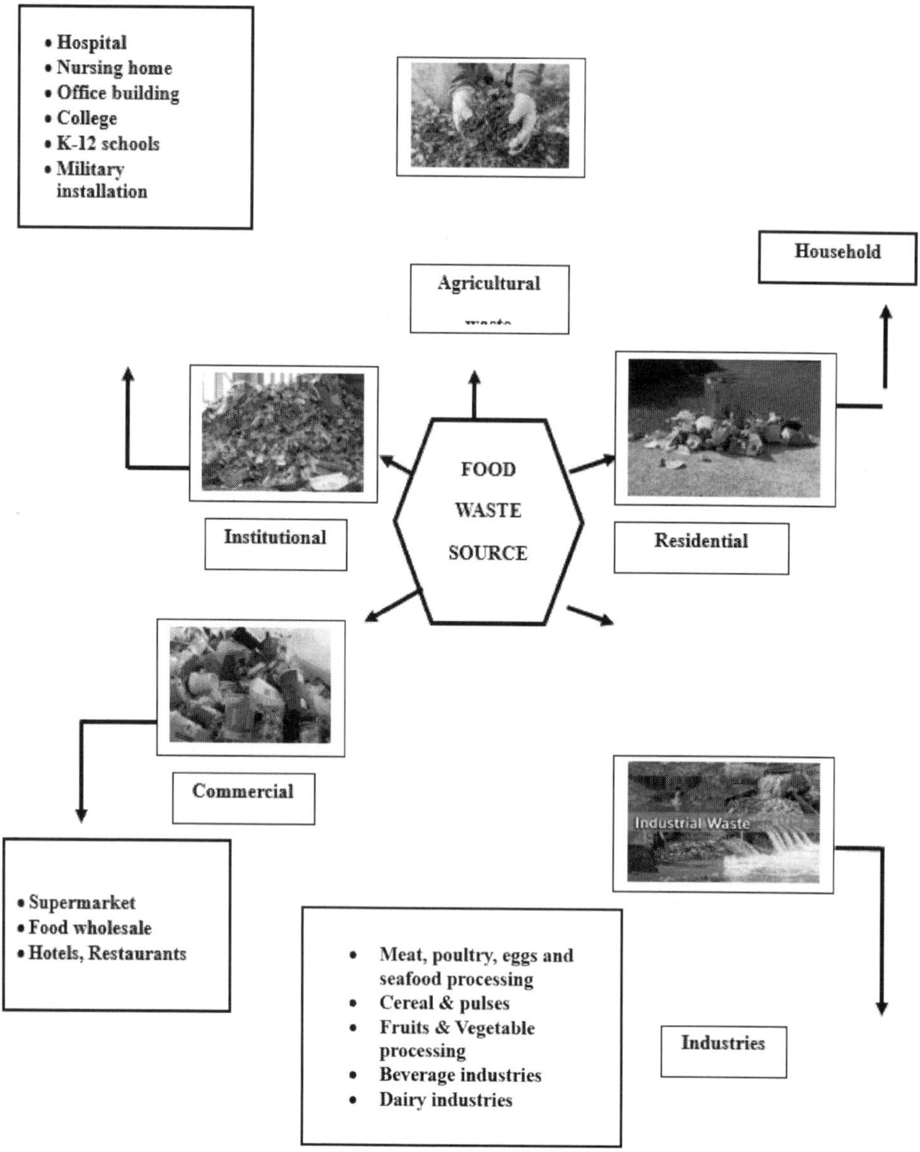

Fig.1: Different sources of food waste

Sources of Food Waste

The FAO estimated about 1.3 billion tonnes of wasted food are generated annually which is 1/3rd of total food production on global scale (Gustavasson et al., 2011). Food waste also leads to the losses of other resources like water, land, manpower and energy. There are several sources such as food processing industries, agricultural waste for the generation of food waste (Sharma et al., 2020). There are many countries of food waste production in India is 63, 68 and 90 kg/capita/year (Grover and Singh 2014).

Environmental Impacts of food waste disposal

When food is wasted rather than eating, the environmental effects of the food production and consumption are further compounded. According to FAO, 1/3rd of all food produced for human consumption worldwide is lost or wasted along the whole supply chain (FAO 2011). In 2019, about 931 million tonnes of food were wasted globally, accounting for the 17% of all food to be consumed (Zhou et al., 2022).

India produces lots of food waste but the techniques for disposing of it are quite inadequate; instead, organic garbage is typically dumped at landfills. Landfills is the most common method for the food waste disposal in developing nations including India, which accounts for the 90% of total food waste (Thi et al., 2015).

Food waste and environmental concern: a global macro problem

Previous studies have shown the positive correlation between greenhouse gas footprint and water footprint with reduction of global food waste (Chen et al., 2020). The recent COVID-19 a global pandemic and result lockdown in the many affected country may serious threaten current waste management policy. Thus, food waste has emerged as a global threat due to the poor management in the global pandemic (Jribi et al., 2020).

Future opportunities and challenges

Sustainable management of food waste include horticultural waste and domestic kitchen waste, through the biorefineries emerged as a potential technique. Establishing an industrial biorefinery for the bioconversion of food waste into valuable compounds is a major challenge related to sustainability. In contrast food waste recycling was not separated but include municipal waste in developing countries. Poor food waste management can create several environment pollutants, including green house gas and toxic environmental contaminants. Although the thermal processing of food waste and biochemical processing is a well adopted valorization or the management process to tackle industrial waste food.

Recently the efficiency of food waste as an adsorbent for removal of the hazardous dye from wastewater has been examined (Sridhar et al., 2022).

Considering as food waste is a serious cause of concern. In this topic we have covered the concern, management strategies, challenges and future perspectives regarding food waste management in India.

Food waste in india mainly generated through the domestic, commercial, agricultural and industrial sources. Like other countries, India may take certain steps to reduce the food waste like with collaboration by the charities and food banks to make sure that surplus food from the stores is given to people need. There are many cases in the point of research that aims to regulate food waste sustainably. This also discusses on the environmental impact of the food waste disposal. Improper food waste disposal practices have several negative environmental effects such as pollution, emission of GHGs. Food waste valorization is a promising sustainable approach for addressing global food waste and its conversion into value added products. It is generally derived from the domestic or industrial waste streams. As a result, food waste has become a global concern because of the unplanned management of the other solid waste. The integration of the biorefinery has been promising option for the industrial food waste management. Domestic and kitchen wastes should be managed separately with the initiative of waste separation from municipal solid waste. However several studies have shown diversified food waste is a potential and untapped source of the bioactive molecules and bioenergy.

Cultural Attitudes towards Waste Perceptions of waste in different cultures Overcoming stigma and changing mindsets

Food waste remain to be a challenge for the global efforts that promotes the sustainable consumption and include severe social, environmental and economic consequences (Aka and Buyukdag 2021). The economic cost of the loss of the food products and wastes equates to 1 trillion (USD) globally and account for 8% of the total global greenhouse emission (FAO 2023). A significant amount or proportion of wasted foods is perfectly edible and is discharged for being unattractive in the appearance due to the proximity to expiry or the packaging damage (Grewal et al., 2019). Previous study has categorized this type of food waste as a sub-optimal or the imperfect food waste, which is defined as avoidable food waste caused by ill-informed assumptions that devalued the edible foods, driven by the lack of familiarity of food and knowledge, food systems and origins (White et al., 2016). The FAO study used compiled global food production and waste data from 2009 estimated that 32% of all food which is produced for the human consumption globally is lost (FAO 2021). Food waste is well known in the challenge in the modern society, either as a result of spoilage or due to consumer's mindset that food past their best by date should be discarded (Traunt 2015).

Promoting acceptance of ugly produce defined as marketing imperfect produce or the irregular produce by normalizing consumption of the food waste recovered annually (Briggs et al., 2016). Food waste is becoming increasingly viewed as a social irresponsible, and the retail ugly foods management is quickly falling behind modern consumer expectations, damaging image of retailers (Grewal et al., 2019).

There are numerous studies that have investigated food waste causes, impacts and solutions (Parfitt et al., 2010). Another categorization refers to avoidable, possibly avoidable and unavoidable food waste depending on the possibility of reusing food material (Johnson et al., 2009). This model defines six characteristics which differentiates in national cultures and the behavior, belief and life approach of individuals (Venaik et al., 2013).

Significance of Food Wastes from Nutritional, Environmental, and Economical Points of View:

The food waste is considered as a source of complex carbohydrate, protein, lipids, and etc. therefore potential health benefits of the wasted food rely on their high content of biologically active compounds (Jaiswal 2016). Food waste has social impacts on the associated with nutrient loss and world hunger. The wasted foods can theoretically fill nutritional gaps for millions of people. Food waste has environmental and economic impacts right along with its social impacts. Food waste has devastating effects on climate change.

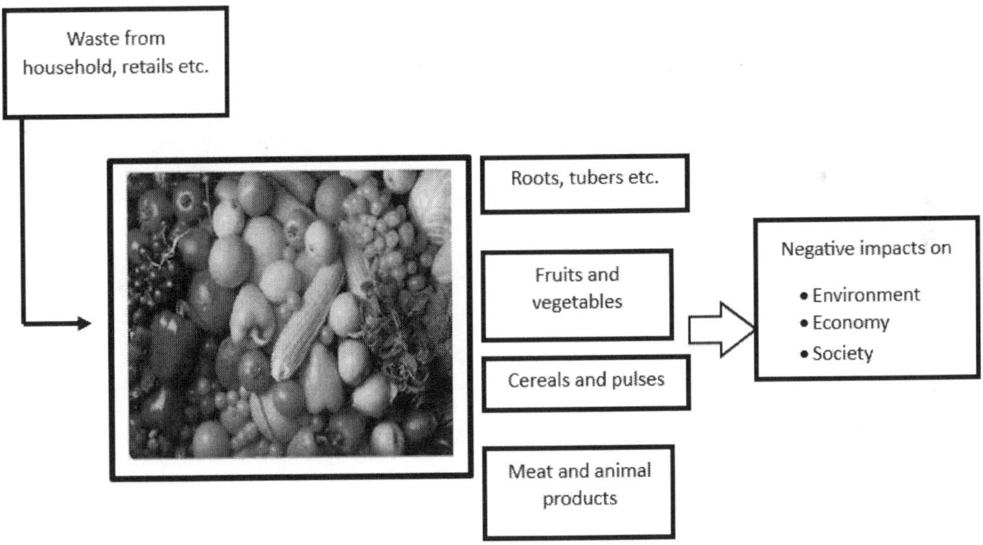

Fig: Impacts of Agricultural food waste and by products

Methodology

The objective of this is to determine the impact of the national culture on food wasting pattern of the consumer's fruit and vegetables. In order to test if the national culture has an impact on waste of food, whether any of the cultural characteristics of a society according to Hofstede's cultural dimensions model has impact on waste foods (Hofstede et al., 2010). Based on this information, following hypotheses were developed:

Hypotheses 1 (H1): The quantity of food waste of fruits is influenced by the characteristics of the national culture.

Hypotheses 2 (H2): The quantity of food waste of vegetables is influenced by the characteristics of the national culture.

Novel Approaches for the Valorization of Agricultural Food Wastes/ Byproducts

The value-added products, which are fine chemicals, nutraceuticals, antioxidants, bio-actives, bio- polymers, bio-peptides, antibiotics, industrial enzymes, bio-nanocomposites, single-cell proteins, polysaccharides, activated carbon adsorbent, chitosan, corrosion inhibitors, organic acids, pigments, sugars, wax esters, and xanthan gum, can be recovered by using food wastes as a substrate (Kavitha et al., 2020). The conventional treatment of the food waste such as landfilling and incineration leads to environmental, economic, and social problems. Special chemicals refined from food waste, varying from solvent to antioxidant materials, are essential for nutraceutical and biomaterial applications.

Consumer Food Waste Perceptions, Knowledge and Behaviors

Participants reported that variably wasting different types of food with the least wasted food item being packed foods (60%) followed by the milk. Approx. half 52% of the participants reported that they rarely or never throw away the leftover. Knowledge regarding the amount of the food waste in US with almost 1/3rd of participants.

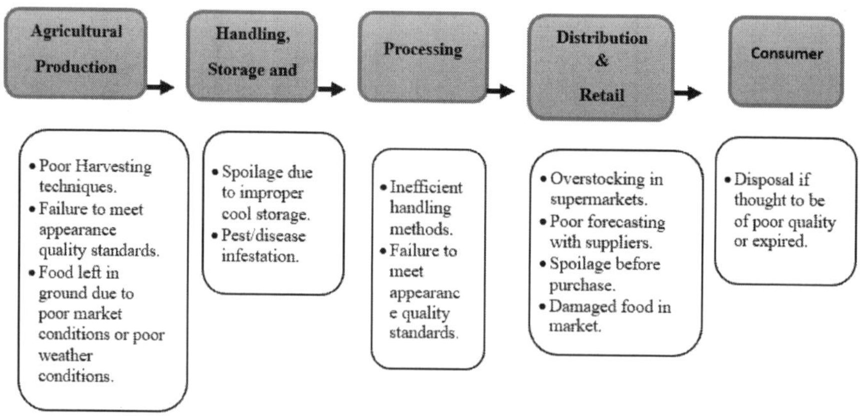

Figure - Causes of Food Loss and Waste in the Food Supply Chain

Discussion

The results of this research confirm the fact that national culture has an impact on the quantity of wasted fruit and vegetables. Consumers with a higher uncertainty avoidance index and higher indulgence index have the tendency to waste more fruit and vegetables. Countries with a higher uncertainty avoidance index have the tendency to waste more food, both of fruits and vegetables product categories. Another categorization refers to avoidable, possibly avoidable and unavoidable food waste depending on the possibility of reusing food material. The high wasted quantities can be also explained by a bad management of the existing production. The indulgence index has a catalyzing effect. They increase the effect of the uncertainty avoidance, but they do not have a significant influence of their own.

Conclusion

The complex reasons behind why nearly 1/3rd of all food produced for human consumption is wasted are evidence throughout food supply chain, from production to consumption. The largest barrier to eliminate the food loss and waste is the corporate control of global food system. The most commonly reported reasons for wasting food were for forgetting about food and disposing it due to safety concerns. However, studies noted they are willing to reduce food waste, the majority reported that they did not seek out information about strategies to reduce household food waste. The food waste campaigns should be administered through local community organizations and enterprises to provide households with strategies to reduce food waste including tips with meal planning and safe foods storage. The largest barrier to eliminating food loss and waste is the corporate control. A large percentage of food waste today is due to microbial or fungal spoilage. Although the established preservation techniques have tackled this issue, consumers increasingly desire more natural forms of food preservatives.

Plants are natural reservoir of antimicrobial compounds and can be exploited to reduce the food spoilage in a variety of ways. In order to diminish the food waste it should adopt several regulations regarding food safety and food loss prevention. The European Union already set a series of sustainable goals, including directive for the reduction of a food waste. In order to reduce the quantities of wasted foods it is important for both the researchers to match the certain food wasting patterns with consumers or the individuals characteristics and the profiles.

7. 3 Gender and Waste Valorization

As the globe faces an urgent need for sustainable waste management, waste valorization has emerged as a potent tool for repurposing waste materials into valuable resources, reducing landfill usage, and mitigating environmental effect. However, the success of waste valorization efforts is dependent on inclusive methods that acknowledge the responsibilities of many socioeconomic groups. Gender, particularly the role of women, is important in waste management, especially in low- and middle-income nations, where women are frequently involved in garbage collecting, sorting, and recycling. (JO Amoah et al., 2023). Exploring how equitable waste management systems might enhance social outcomes and environmental sustainability requires examining the relationship between gender and waste valorization.

Several examples from around the world demonstrate the good impact of including gender into trash valorization. In Brazil, the Women's Movement of Recyclable Material Collectors (MMRC) has been a driving factor in changing trash management methods. This group provides women garbage pickers with the skills, equipment, and resources they need to operate in safer and more equitable settings. By uniting garbage collectors into cooperatives, the MMRC has helped to raise pay, improve working conditions, and promote women as waste management leaders.

Other example in Kenya, women's cooperatives convert organic waste into biogas and compost, which they use to power homes and farms. These cooperatives give women access to technology and finance, allowing them to create useful goods from trash. Women are not just decreasing trash via these projects, but also promoting renewable energy and food security in their communities.

Empowering women in waste valorization offers several advantages for communities, economies, and the environment. First, gender-inclusive waste management has been proven to improve economic benefits. Women who obtain training and have access to financial resources are better able to start small enterprises in recycling and repurposing, which benefits local economies and creates jobs. Gender-sensitive trash valorization efforts enable women to go beyond informal rubbish picking and into skilled employment in the value chain, resulting in higher earnings and more financial stability.

Environmentally, engaging women in waste valorization can boost the effectiveness of recycling and resource recovery. Women are more likely to embrace ecologically friendly trash methods and encourage sustainable waste habits in their communities. When women take on leadership roles in waste valorization, they contribute unique views and information to communities, assisting them in adopting policies that minimize landfill trash, save resources, and cut carbon emissions (MN Muhammad et al., 2013). Inclusive waste management efforts that enable women to meaningfully engage in waste valorization help to achieve the Sustainable Development Goals (SDGs), notably those relating to gender equality, poverty reduction, and environmental protection.

Role of Women in Waste Valorization

In our global society, women's views and attitudes on the environment differ slightly from men's in various societies. Women place more importance on preserving and enhancing nature's capabilities, upholding agricultural lands, and looking out for the future of the environment and nature. Women are the backbone of community-level garbage management in many low- and middle-income nations. They carry out a variety of tasks, including as gathering and classifying waste as well as recycling and material reuse. Women usually labor in the informal sector, where they face less job security, lower pay, and restricted access to resources, whereas men often hold the majority of positions in established waste management systems. Nevertheless, by encouraging recycling, upcycling, and lowering landfill waste, they make a substantial contribution to environmental sustainability.

Women frequently have in-depth, practical knowledge of waste materials and sustainable techniques due to their close engagement in everyday trash-related duties. Indeed, studies show that women are more likely than males to support and implement sustainable waste management practices in the home. (GM Adebo et al., 2012). Societies may improve trash valorization's efficacy and build a more sustainable, inclusive future by acknowledging women's unofficial contributions and incorporating them into official waste management systems.

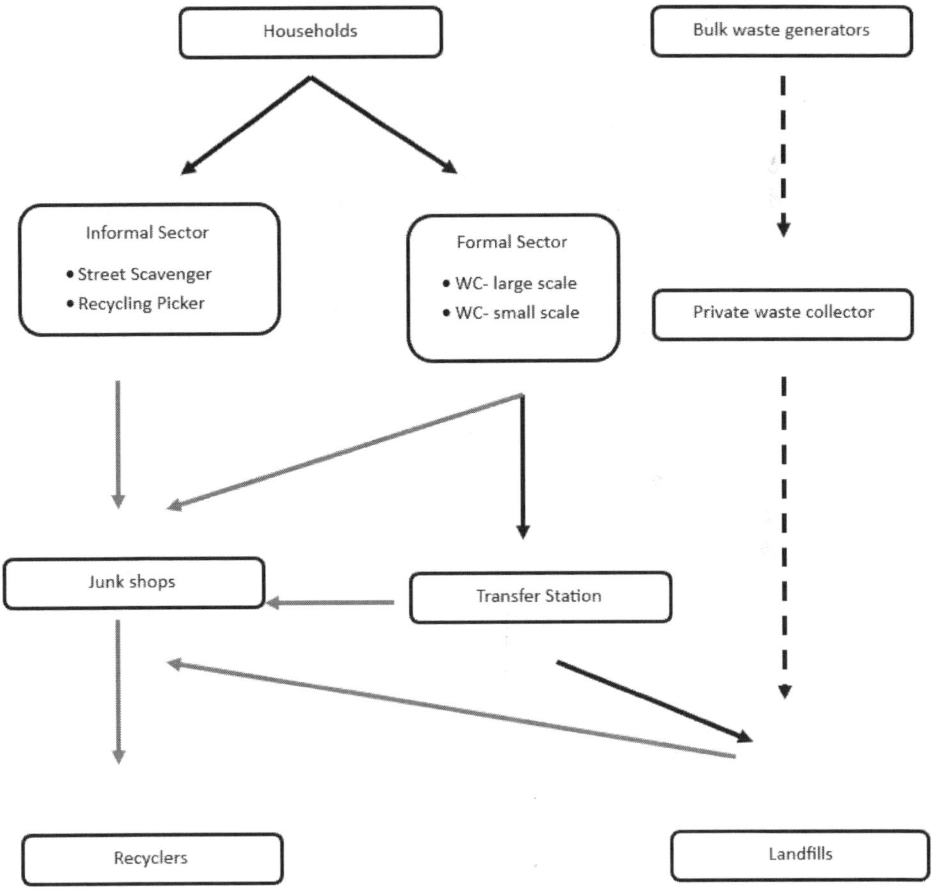

Fig 1: The Role of Gender in Waste Management

There are several advantages for communities, economies, and the environment when women are empowered in waste valorization. First, women can reduce poverty and promote economic growth by converting garbage into useful goods when they have access to capital, technology, and training. Additionally, since they use the networks and expertise that women contribute to these endeavors, gender-inclusive trash valorization programs typically have greater success rates. By lowering poverty and generating jobs, especially for underserved populations, empowering women may help increase social fairness.

Additionally, gender-inclusive garbage valuation benefits the ecosystem. Research indicates that women are more inclined to embrace eco-friendly practices and support recycling and sustainability initiatives at the local level (HI Manu et al.,2013). Societies may capitalize on these inclinations to encourage trash reduction, foster sustainable innovation, and push wider acceptance of recycling techniques by empowering women in waste management and providing them with leadership positions.

Gender-Specific challenges and opportunities

Social conventions, financial constraints, and environmental requirements interact intricately to provide gender-specific waste valorization possibilities and obstacles. The process of turning waste materials into energy or useful goods, or waste valorization, is becoming more and more important for circular economy growth and sustainable waste management. Even though women are frequently the major waste managers in households and communities, particularly in low- and middle-income nations, the importance of gender in this area has long been disregarded. Developing a fair, efficient, and sustainable waste management system requires acknowledging and resolving the particular difficulties women have in trash valorization as well as utilizing the possibilities that are accessible to them.

Women have a large and diverse role in trash management and valuation. The management of household garbage, including waste segregation, recycling, and disposal, falls on women in many societies. Women frequently gather, sift, and reuse waste items, including plastics, metals, and organic garbage, to produce secondary goods or boost local economies through recycling networks in the informal sector. Although they frequently labor under difficult conditions, such as unfavorable working conditions, low pay, little job security, and health risks, these positions make a substantial contribution to the reduction of trash and environmental damage (OC Ajewole et al., 2012).

Women's access to resources, opportunities, and rights is restricted by larger cultural norms, which can cause gender-specific difficulties in waste valorization. Lack of access to education and training, especially in technical and management abilities, is a major problem. In vocational programs that teach sustainable resource management, recycling technology, and trash processing, women are often underrepresented. Due to their lack of training, they are frequently limited to the more difficult, low-paying jobs of garbage collecting and sorting, which hinders their advancement in the waste valorization industry. In the waste management value chain, this means that women lose out on prospects for higher-paying, specialized jobs in areas like material processing, plant operation, or business ownership.

Not with standing these obstacles, there are plenty of chances to promote gender parity in garbage valuation, which may be advantageous for economies, communities, and the environment. Policymakers and organizations may unleash a number of benefits, such as better waste management results, economic empowerment for women, and environmental sustainability, by establishing supportive frameworks for women in trash valorization. Creating focused education and training initiatives that give women technical know-how in recycling, material recovery, and sustainable resource management is one such possibility. By closing the gender gap in technical knowledge, these programs can help women transition into specialized fields with better pay and more stable employment. Additionally, training programs give women the business skills they need to launch and run waste-based businesses, boosting local economies and creating jobs.

Another important area where focused actions can have an impact is financial inclusion. Government subsidies, microfinance programs, and modest grants can provide women the money they need to invest in trash valorization projects. Women who have access to financial resources can buy the equipment and supplies needed for effective trash processing, increasing their output and income (GM Adebo et al., 2012). Financial initiatives that provide flexible repayment plans or low-interest loans can also motivate women to grow their waste management companies, which might lead to the creation of jobs for other local women. By attracting more participants, financial inclusion not only improves the economic security of women but also fortifies regional waste valorization systems.

Policymakers, organizations, and communities must collaborate to eliminate the structural hurdles that prevent women from participating in trash valorization in order to develop a truly gender-inclusive strategy. This entails passing laws that give all employees in the waste valorization industry social protections, safe working conditions, and equitable compensation. Establishing a safe and courteous work environment for women requires both gender-sensitive health and safety regulations and legal rights against workplace harassment. Furthermore, data collecting on trash valorization and gender-focused research may be utilized to track the effects of gender-inclusive policies and pinpoint particular difficulties faced by women.

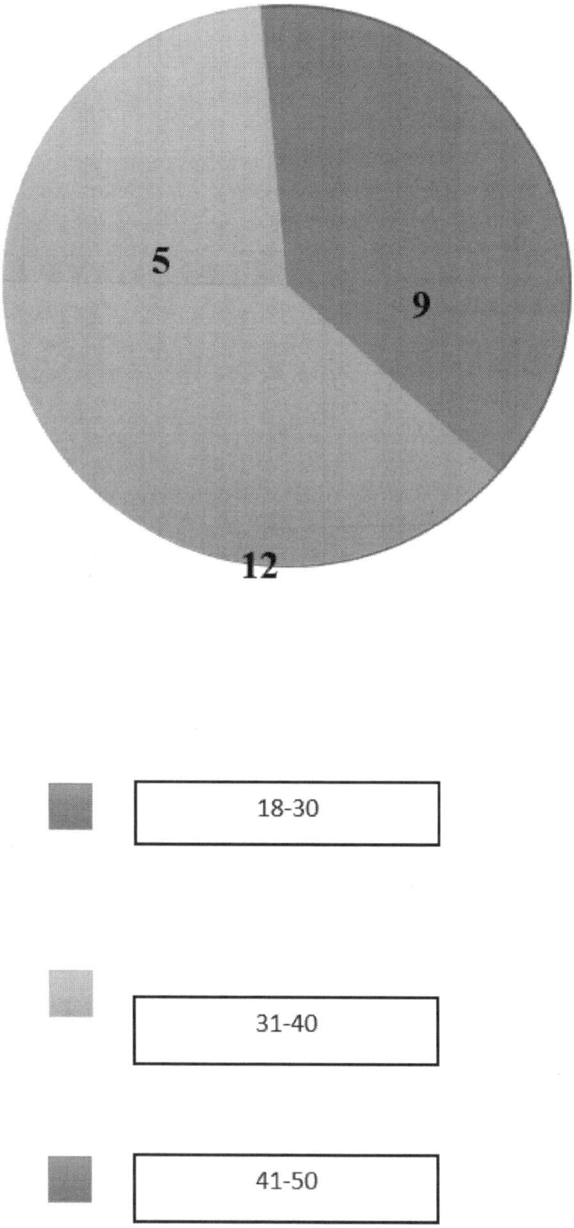

Fig 2: Age-group distribution of women workers

Education and Awareness Programs Training and capacity-building initiatives Success stories of community-led valorization projects

The food waste is a problem that have become a major topic of concern for the international institutions, such as the food and agricultural organizations (FAO), with an extensive impact at the social and environmental levels. Reducing the food waste is correlated with the 2nd and 12th Sustainable Development Goals of the United Nations: "Zero Hunger" and "Responsible Consumption and Productions", respectively. The European Union is committed to reduce the food waste by half by 2030 and, as a part of its strategy, it established the EU Platform on Food Losses and Food Waste in 2016, where prevention measures can be defined as, best practices can be shared, and progress can be evaluated over time. Reducing food waste has also been placed on the agenda of national policymakers, within a program such as the "Waste Prevention" program of German government, the "Stop Food Waste" program in Ireland (FAO 2013).

Food waste occurs along entire supply chain, in the earlier stage, in the case of developing countries, and in the next stages, at a consumer level, within developed countries (Gustavsson et al., 2011). Various food waste mitigation measures can be used as a food wastage pyramid, from the most to the least environmentally friendly measures: reduction, re-use, recycle, recovery, and disposal (FAO 2013).

Population is also concentrated unevenly, exposing a citizen in some countries to natural hazards more than citizens in others. This, in turn, causes a variety of awareness levels among individual and society, and these differences influence disaster reduction measures and their efficiency. Increasing hazard awareness is the only one way to reduce the negative impacts of natural disasters, as the people aware of risks are more likely to reduce potential losses (Dunbar 2007).

Disasters may become more frequent and severe in future because of the population growth, climate change and migration to the areas under a greater risk, and people increased reliability on infrastructure (Stocker et al., 2013). Prior to a research in educational approaches to teaching behavioral action suggests that teachers should not begin by teaching action related to specific natural hazards, but by constructing a knowledge base from bottom-up to develop informed response (Oyao et al., 2015).

Environmental Education Collection For Sustainable Development

The main result regarding to a macro-education is the launch of the Environmental Education Collection for Sustainable Development, with a seven volumes (today in the third edition). This kind of materials serves as a primary and secondary education and is used for the training of moderators, community leaders and a public and private organizations, making them able to jointly develop their own socio-environmental responsibility methodologies (Brasil 2012).

Instruments And Options For Promoting Bes

Environmental governance is facilitated by a diverse institutional arrangement, ranging from market tools and community-based approach, involving private and civil society actors or combination of these, along with the underlying principles of governmental command and control mechanisms (Muradian and Rival 2013). Economic instruments, such as taxes or tradable permits and quotas, can complement regulation in place as well as public engagement-based approaches.

The agro-food system and the inclusion of BES as "produced" goods and services, we focus on the role of economic instruments that use certain degree of price signals and awareness raising options to promote environmental change (Hahn et al., 2015).

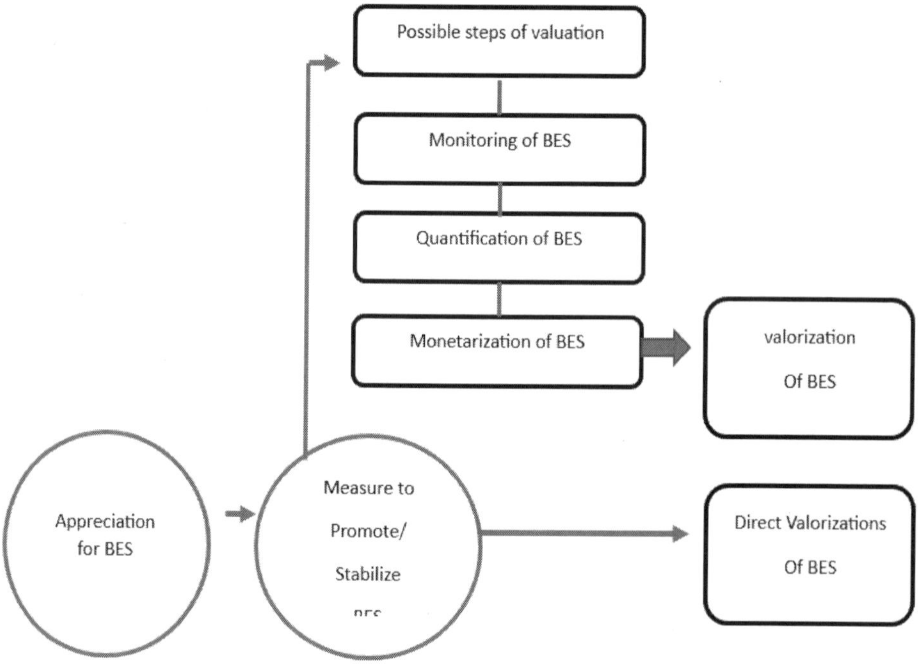

Fig 1: valorizations Process

In this study, the relationships between the level of education and components of the WRI were analyzed in order to create a quantitative basis for evaluation of education's role in enriching knowledge on natural hazards and thereby raising awareness and consequences mitigation. The study showed that education factor calculated independently of WRI methodology was significantly correlated with several components of WRI: Natural Hazard Factor or relative vulnerability, susceptibility, lack of adaptive capacities. Therefore, further investment in education, especially in countries suffering from poverty and social problems, was seen as crucial in lowering the WRI and supporting natural hazard awareness.

CHAPTER 9

POLICY AND REGULATORY FRAME WORK

Global and National Policies on Waste Management

Overview of Existing Regulations and Policies

Good waste management is essential to sustainable development as improper disposal leads to environmental degradation, health risks, and economic losses. There are different global regulations and policies that strive to standardize the way waste is handled and promote sustainability. For example, the European Union's Waste Framework Directive outlines a legal framework for waste management in member states, highlighting prevention, reuse, and recycling (European Commission, 2020). At national levels, Japan and Germany have implemented tough regulations embracing circular economy principles, making landfills almost redundant (Smith et al., 2018).

International Agreements and Protocols on Waste Management

International agreements have been central in harmonizing efforts across borders. The Basel Convention on the Control of Transboundary Movements of Hazardous Wastes and Their Disposal is one of the landmark treaties in terms of reducing the generation of hazardous wastes and their sound management environmentally (Basel Convention, 2019). On the same level, the Stockholm Convention seeks to eliminate persistent organic pollutants, most of which are byproducts of waste, in protecting human health and the environment (UNEP, 2021). Such frameworks are supplemented by regional agreements such as the Bamako Convention in Africa, whose focus is on the proper management of hazardous wastes on the continent (Bamako Convention, 2017).

Waste Valorization through Policy

Incentives, Subsidies, and Support Mechanisms:

Across the world, governments have been motivated by the economic and environmental benefits of waste valorization-the practice of converting waste into value-added products. Such policy instruments include tax incentives, grants, and subsidies. For example, South Korea's Extended Producer Responsibility (EPR) program mandates manufacturers to manage the lifecycle of their products, encouraging waste reduction and valorization (Lee et al., 2019). In the European Union, subsidies for biogas production from organic waste have boosted renewable energy generation and reduced landfill use (European Biogas Association, 2021).

Regulatory Challenges and Gaps

Despite all these developments, enormous regulatory gaps are preventing waste valorization from reaching its full potential. In most developing countries, policies are ambiguous, and the mechanisms of enforcing them are inadequate (Swapnil et al., 2020). There are also no harmonized standards for recycled materials, which create barriers to international trade and market acceptance (World Bank, 2018). Such a situation requires collaboration from the government, private sectors, and international organizations.

Case Studies of Successful Policy Interventions

Examples from Countries with Advanced Waste Valorization Systems

Sweden and Singapore are examples of good waste valorization policies. Sweden has been able to reduce landfill rates to less than 1% of the total waste through its effective waste-to-energy initiatives and strong policy frameworks (Johansson et al., 2017). Singapore has also implemented policies that promote recycling and incineration with energy recovery, with a recycling rate of 60% by the National Environment Agency (NEA, 2020).

Lessons Learned and Best Practices

Key lessons from these countries include public awareness campaigns, robust infrastructure, and stakeholder collaboration. For instance, Sweden's success in household waste segregation is due to sustained public education efforts (Johansson et al., 2017). Singapore's integrated waste management system underscores the value of aligning economic incentives with environmental goals (NEA, 2020).

Future Policy Directions

Recommendations for Improving Policy Frameworks

To improve waste management and valorization, governments should focus on integrated policies that incorporate economic, environmental, and social goals. Recommendations are as follows:

- EPR programs need to be strengthened to make producers accountable for the entire product lifecycle (Swapnil et al., 2020).
- Global regulations for recycled materials should be standardized to promote trade and acceptance (World Bank, 2018).
- Investments in research and innovation should be increased to develop advanced technologies for waste valorization (UNEP, 2021).

Role of Governments, NGOs, and International Organizations

Governments play a central role in policy formulation and implementation, and NGOs and international organizations can play an important role by providing support through advocacy, capacity building, and funding. For example, the United Nations Environment Programme (UNEP) provides technical support and policy advice to member states (UNEP, 2021). Such collaboration can help accelerate progress toward a circular economy.

CHAPTER 10

CASE STUDIES ON SUCCESSFUL WASTE VALORIZATION

The process of turning agricultural waste into useful products, or waste valorization, reduces dependency on non-renewable resources while addressing pollution. This strategy involves turning agricultural waste into high-value goods like biofuels and bioactive chemicals, which improve economic sustainability while simultaneously reducing negative environmental effects. Examples of how these technologies are being used to generate sustainable energy solutions, support rural development, and advance circular economy principles include the production of biogas and bioethanol from wastes in Germany and India. Both nations want to lessen the negative environmental effects of traditional disposal techniques, such as open burning and landfill use, by repurposing agricultural waste (ESPON report, 2020) (Urban vacancy in Europe). Valorization of agricultural waste is an important process that turns agricultural wastes into useful products, solving waste disposal-related environmental issues while also promoting economic growth. Utilizing agricultural waste, including crop leftovers and livestock manure, new technologies can produce biofuels, bioproducts, and bioactive substances (Kumar et al., 2021; Gupta et al., 2022). The creation of biogas from animal waste in Germany and the value-adding of crop leftovers in India are two noteworthy case studies that are presented in this chapter. By exploring the history, technological approaches, financial effects, and important takeaways, each case study will demonstrate the revolutionary potential of agricultural waste.

Case Study 1 – Crop Residue Valorization in India

India's agricultural sector generates large volumes of crop residues, such as rice husks and sugarcane bagasse, which are often disposed of through open burning, causing air pollution and wasting potential biomass resources. Recent valorization projects in India have focused on converting these residues into biochar and bioethanol. India, one of the largest agricultural producers globally, generates approximately 500 million tons of crop residues annually, predominantly from rice, wheat, and sugarcane cultivation (Singh et al., 2022). Traditionally, these residues have been disposed of through open burning, a practice that not only contributes to severe air pollution but also increases greenhouse gas emissions (Bhaduri et al., 2020). To combat these environmental issues, various government initiatives and research projects have been introduced to valorize crop residues, specifically focusing on rice husks and sugarcane bagasse for conversion into biochar and bioethanol, respectively (Kumar et al., 2021; Sharma et al., 2023).

Biochar from Rice Husks: Using mobile pyrolysis units, rice husks are converted into biochar, which enhances soil health and crop productivity when applied to farmland. Biochar increases soil organic content, improves water retention, and reduces emissions by substituting for the traditional practice of open burning. Biochar's soil benefits, including carbon sequestration and increased nutrient availability, have been well-documented in sustainable agricultural practices, contributing to India's initiatives for soil health management (Chen et al., 2019)(Urban vacancy in Europe). A noteworthy project aimed at the valorization of rice husks involved converting these residues into biochar using mobile pyrolysis units, which could be transported and operated in diverse agricultural regions (Choudhary et al., 2021). This innovative approach not only minimized the logistics costs associated with transporting the residues but also significantly reduced emissions linked to conventional burning practices (Tripathi & Sharma, 2023). Farmers were trained to operate these pyrolysis units, and the biochar produced was subsequently applied to improve soil fertility and enhance carbon sequestration (Verma et al., 2022).

Bioethanol from Sugarcane Bagasse: Advanced fermentation techniques enable the transformation of sugarcane bagasse into bioethanol, contributing to rural energy independence and creating employment opportunities. This aligns with India's energy sustainability goals by reducing fossil fuel dependency, providing an environmentally friendly and economically viable alternative (Al-Hamamre et al., 2017)(Urban vacancy in Europe). Additionally, the valorization of bagasse for bioethanol production has catalyzed local job creation in rural regions, supporting both energy and economic sustainability. The bioethanol produced from sugarcane bagasse offered a renewable energy source, significantly diminishing reliance on fossil fuels. The economic analyses revealed encouraging returns on investment, attracting private investments and promoting the development of biorefinery infrastructure in India. For instance, one biorefinery project reported an annual production capacity of 10 million liters of bioethanol, generating substantial revenue and creating around 150 local jobs (Patel et al., 2020). Additionally, this initiative spurred rural development by creating job opportunities in processing plants and increasing farmers' incomes through sales of bagasse (Singh et al., 2022

Case Study 2 – Biogas Production from Animal Waste in Germany

In Germany, livestock waste is commonly valorized through biogas production, an integral component of rural energy security and waste management efforts. Biogas production transforms animal waste into renewable energy, which significantly reduces the environmental impact associated with waste disposal. Germany's livestock industry produces significant quantities of animal waste, posing considerable environmental risks if not managed properly (Müller et al., 2021). In response to these challenges, the country has implemented numerous biogas projects aimed at converting animal waste into biogas, aligning with national renewable energy objectives (Lange et al., 2022). These initiatives have not only improved waste management practices but have also significantly contributed to energy security in rural areas (Schmitz et al., 2020).

- **Biogas Plant Operation:** Through anaerobic digestion, biogas plants convert animal waste into electricity and heat, supporting Germany's renewable energy targets and providing farmers with an effective waste management solution. Research suggests that biogas production reduces methane emissions by capturing and utilizing methane, which would otherwise be released into the atmosphere. This transformation aligns with Germany's policies on renewable energy and carbon reduction (JUST2CE project report, 2023)(Vironen_etal_EPRC_2022_…). The biogas production process entails several key steps. Initially, animal waste is collected and subjected to pretreatment to remove contaminants, making it suitable for anaerobic digestion in sealed digesters (Müller et al., 2021). This biological process effectively converts organic matter into biogas, which primarily comprises methane and carbon dioxide (Lange et al., 2022). The raw biogas is then upgraded to eliminate impurities, making it suitable for use as a natural gas substitute or as fuel for vehicles (Schmitz et al., 2020).
- Economic and Social Impact: Integrating biogas plants within rural communities supports the agricultural economy by creating jobs in plant operation and maintenance. The German government's subsidies for renewable energy projects further incentivize biogas production, contributing to both economic viability and social acceptance. The community-driven approach to biogas has strengthened rural economies while advancing Germany's renewable energy goals (Toppinen et al., 2017) (Vironen_etal_EPRC_2022_…). The biogas production process entails several key steps. Initially, animal waste is collected and subjected to pretreatment to remove contaminants, making it suitable for anaerobic digestion in sealed digesters (Müller et al., 2021). This biological process effectively converts organic matter into biogas, which primarily comprises methane and carbon dioxide (Lange et al., 2022). The raw biogas is then upgraded to eliminate impurities, making it suitable for use as a natural gas substitute or as fuel for vehicles (Schmitz et al., 2020).

- **Market Development and Challenges for Bio-Products in Germany**

The bio-products market in Germany has expanded due to increased demand for sustainable products and supportive policies.

- **Market Growth:** Germany has built a strong market for bio-products like biochar, bioethanol, and bioplastics. Governmental support for sustainable development and consumer awareness of environmental issues have contributed to this growth (WABEF report, 2017)(WABEF final narrative r…). This market also benefits from Germany's commitment to circular economy principles, as bio-products help close resource loops by reintroducing agricultural residues into the economy.
- **Challenges:** The market faces hurdles such as high variability in waste materials, high initial investments, and competition from conventional products. Waste composition variability affects bio-product quality, while high capital investment limits entry into bio-processing. Addressing these challenges will require further innovation to streamline processes and reduce production costs (Thesis_ReaPärli, 2023)(WABEF final narrative r…).

In Germany, community-led initiatives are integral to promoting local economic empowerment, enhancing public health, and raising environmental awareness. Grassroots organizations collaborate with local businesses and municipalities to develop sustainable waste valorization practices. Community-driven projects, particularly those involving biochar and composting, reinforce circular economy principles and strengthen local economies. By managing waste locally, these initiatives encourage environmental stewardship and foster community involvement, crucial for sustainable development (Urban Vacancy in Europe report, 2023)(Urban vacancy in Europe). Community-led efforts in waste management, as observed in various EU case studies, promote a sense of shared responsibility and collective action.

Utilizing local resources and optimizing production processes is essential for lowering costs and making bio-products competitive in the market. Techniques such as decentralized production models and resource-efficient practices help reduce dependency on imported resources and align with the principles of sustainability (Social Vulnerability Report, 2024)(WABEF final narrative r…).

Integrated bio-refineries maximize the use of agricultural waste by producing multiple products, including bioethanol, biogas, and fertilizers, reducing environmental impact. These refineries promote sustainability by fully utilizing waste streams and reducing greenhouse gas emissions from traditional waste management methods (Torrie, 2014)(Vironen_etal_EPRC_2022_…). Collaborations with research institutions, governments, and private investors are essential for advancing bio-product technology and securing funding for scaling operations to commercial levels. Case studies demonstrate that partnerships enhance innovation and market readiness (Vironen et al., 2022)(Vironen_etal_EPRC_2022_…). Growing environmental awareness among consumers has fueled demand for bio-products, facilitating market growth and acceptance. Educational campaigns highlight bio-products' environmental benefits, fostering a supportive consumer base committed to sustainable consumption (ESPON report, 2020)(Urban vacancy in Europe).

Case Study 3: Bio-products from Food Processing Waste in Europe

In Europe, innovative projects are making significant strides in valorizing food processing waste by converting it into high-value bio-products. One notable example is the EU-funded Agrimax project, which aims to maximize the value of agricultural and food processing waste by developing a range of bio-based products. The project has established pilot plants in Italy and Spain to process these wastes into bio-polymers, bio-composites, bio-based coatings, and active packaging for the food, packaging, and agricultural sectors(d0ad045d-en). The Agrimax project specifically targets by-products like tomato pomace, olive pomace, and potato peels, converting these into sustainable products such as bioplastics and bio-fertilizers. These materials can serve various industries by replacing conventional, petroleum-based materials, thus contributing to a circular economy. Furthermore, the use of bio-fertilizers derived from these processes provides an eco-friendly alternative to synthetic fertilizers, promoting soil health and reducing chemical run-off into water bodies(d0ad045d-en).

Another significant initiative, the Agrocycle project, addresses the recycling and valorization of waste streams from the agro-food sector by providing guidelines through the AgroCycle Protocol. This blueprint outlines pathways for sustainable agro-food waste valorization, targeting the transformation of organic residues into specialty chemicals and biofuels. By following this protocol, industries can repurpose agricultural residues into bio-based chemicals, which are essential for producing biodegradable packaging and other environmentally sustainable products(d0ad045d-en). These projects not only offer environmental benefits by reducing waste but also enhance economic opportunities by creating new bio-based products that are gaining popularity in European markets. The high-value compounds extracted from waste streams, such as natural additives, flavorings, and microbial growth media, underscore the versatility of food processing waste as a resource for innovative applications in the bio-economy(d0ad045d-en).

Through these efforts, Europe is advancing its bio-economy and circular economy goals, demonstrating the commercial viability of using food processing waste as a renewable resource for creating high-value bio-products. These case studies highlight the importance of research and development in achieving the sustainability goals set by the European Union and fostering a market for bio-based alternatives to traditional materials(d0ad045d-en). This analysis exemplifies how food processing waste in Europe is being transformed into sustainable solutions that cater to diverse industries, reinforcing the role of bio-refineries in supporting regional and global sustainability targets.

Case Study 4: Community-Led Waste Valorization in Germany

Germany has made substantial progress in implementing decentralized and community-driven waste valorization initiatives. These projects prioritize the local recycling of organic waste, focusing on the production of biogas and compost from community-collected biowaste. Such efforts align with Germany's circular economy principles, aimed at reducing environmental impact and optimizing resource recovery.

Decentralized Biogas Production :

Decentralized biogas plants, such as the systems promoted by the DECISIVE project, are pivotal in transforming local organic waste into biogas, which provides renewable energy and contributes to waste reduction. These small-scale anaerobic digestion (mAD) facilities are strategically placed within communities to facilitate localized energy production and minimize waste transport requirements. Biogas produced in these systems can be utilized for heating, electricity, or as fuel, while the remaining digestate serves as a high-quality biofertilizer for local agricultural fieldsships and Composting Initiatives

Local municipalities in Germany are also implementing community-led composting systems to process food and green waste. Composting programs are often supported by community education efforts that encourage household biowaste separation, reducing contamination and improving compost quality. For instance, in certain regions, local governments offer resources such as compost bins and workshops on composting best practices, thus fostering a sense of community responsibility and environmental stewardship .

Policy Support any's waste management policies mandate biowaste separation and support community-led waste valorization. The government's regulation (KrWG, § 11 para. 1) on biowaste, effective since January 2015, requires separate collection of organic waste to streamline processing in biogas and composting facilities. These regulations are bolstered by community participation, with households actively engaging in biowaste collection and sorting initiatives. This policy framework promotes the widespread adoption of biowaste recycling and reinforces the role of community-driven projects in achieving Germany's environmental objectives .

Environmental and Economic Impact :

The emmunity-led waste valorization in Germany include reduced greenhouse gas emissions through decreased landfill use, reduced reliance on synthetic fertilizers, and enhanced soil health from compost and digestate application. Economically, these initiatives create jobs in waste management, biogas plant operations, and compost production, supporting the local economy while fostering a circular waste management approach that transforms waste into valuable resources .

Lessons from Case Studies: Key Takeaways and Applicability to Other Regions

The cases of India, Germany, and broader European initiatives reveal several key takeaways that can inform waste valorization strategies in other regions:

1. **Local Adaptation and Decentralization:** Decentralized systems, as seen in Germany, help reduce transportation costs, manage waste locally, and foster community involvement. Local adaptation of valorization strategies, such as India's use of crop residues for bioethanol, ensures relevance to regional agricultural practices and waste profiles(D3.5_Survey on waste co...).
2. **Collaborative Models and Public Engagement:** Community participation, combined with support from local governments, as in Germany, enables smoother implementation of waste valorization programs. Public engagement through educational programs enhances recycling and composting rates and fosters an understanding of waste-to-energy processes(D3.5_Survey on waste co...).
3. **Circular Economy and Integrated Bio-refineries:** Successful bio-products markets, like those in Europe, are often supported by integrated bio-refineries, which maximize resource use by producing multiple products. Cascading uses of biomass, as encouraged by the EU's circular economy strategies, allow for higher resource efficiency and long-term sustainability(d0ad045d-en).

Conclusion

Agricultural waste valorization represents a transformative approach to environmental management, enabling the conversion of organic residues into valuable products while promoting a circular economy. The cases of India and Germany underscore the significance of regional adaptation, community participation, and government support. By leveraging locally available resources and advancing bio-refinery technology, waste valorization can address global environmental issues, reduce greenhouse gas emissions, and improve rural economies. With continued innovation and cross-regional collaboration, waste valorization can become a cornerstone of sustainable development worldwide.

CHAPTER 11

FUTURE PERSPECTIVES AND CONCLUSIONS

Emerging Trends in Waste Valorization, New technologies and innovations on the horizon, Potential future applications and markets:

Around the world, the food sectors, yearly, produce the significant amount of food waste and its by-products. In the processing of food nearly about 38% of waste food in produced. Food waste comes from the variety of sources, like by-product, from many breed animals, like carcasses, foods from other sources like sea-foods etc. because of the low biological stability, large nutritional value, high concentration of organic components. The enormous quantity of food waste and its microbial health, waste treatment costs, adding of the cost of the food producers. Nowadays consumers are looking for the natural, organic, or the healthy foods that are safe. The functional foods globally, about $100 Billion in the demand was predicted in 2013 (Capanoglu et al., 2022). The creation of the functional foods requires integration of the particular health-benefits ingredients. Cancer, Diabetes, and the Cardiovascular disease are the main cause of threat to life in the worldwide. The use of this can prevent certain type of disorders.

Microencapsulation and the encapsulation technology defines the technology of packaging solid, liquid, gaseous component in the continuous film as a coating to produce the capsule from micro to millimeter ranges in size, may be valuable and can be a specialized alternative to maintain the bioactive chemicals (Capanoglu et al., 2022). Due to its useful property, it is widely used in many pharmaceuticals and food industries. Due to its quality it encourages easier handling, transportation of bioactive substance and minimize or to prevent release of the core material when come in contact with environment factors such as heat, air, light, etc (Qin 2016).

Microencapsulation Applications In Food Industry: Global Trends

According to the trend of food industry, global issue such as climate change, population ageing, food waste, unfair trade, these are changing the consumer attitude towards purchasing healthy, plant based, sustainable foods. It provides the fragrance and flavors, used in many foods and beverage items with proper texture mixing, enticing aroma release, tasty flavor.

Food flavor can also be microencapsulated for prevention of evaporation, oxidation, heat degradation for prolong shelf life. This property is expected to increase demand for the microencapsulated flavor in the different food applications. The COVID-19 pandemic had a negative effect on many businesses. It also used in many businesses to impart or cover the bitter taste, boost stability, enable individualized drug administration.

There is increase interest in global market size and it is important as the population ages health care is receiving more focus (Ozkan et al., 2019). Microencapsulation is a dependable technology to address problem associated with preparation of functional food and offer additional benefits like controlled or target release, increase stability, dissolution, bioavailability, and low ability. Additionally, food additives are reduced in accordance with clean label and healthy living trends. The flavor dose can be decreased while maintaining perception through flavor boosting via microencapsulation (Rathod et al., 2018). The global market size was estimated to be around USD 10.95 billion in 2021 and it is projected to reach USD 25.99 billion by 2023 (Consoli et al., 2015).

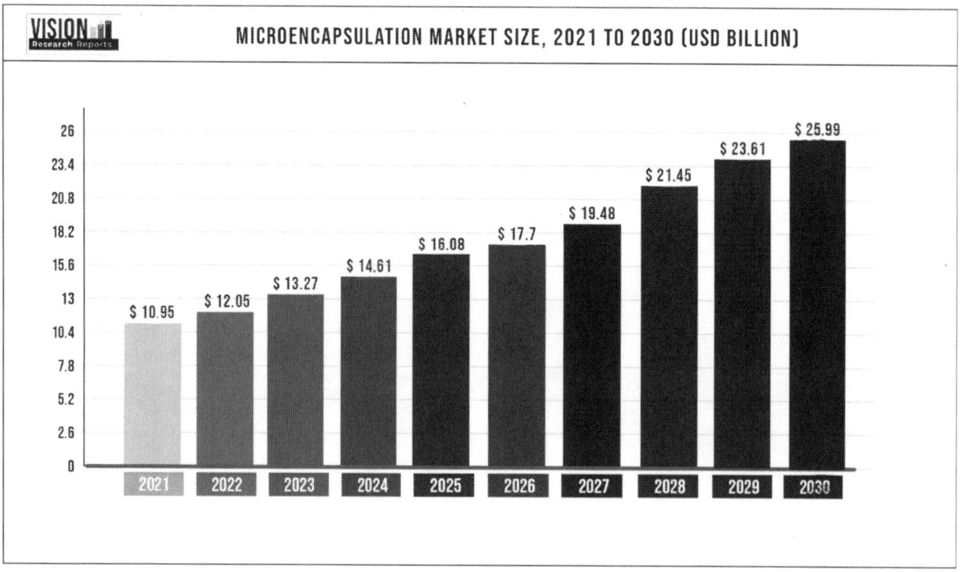

Figure 1. Microencapsulation market size from 2021 to 2030 (USD billion).

Updates In The Valorization Of Food Supply Chain Waste In Different Areas Around The Globe

Cheese Whey Waste Water:

It is a by-product comes from the precipitation and removal of milk casein during the cheese manufacturing. There are 3 types of waste water during cheese manufacturing unit, which herein, referred to as CWW. The organic content of CWW accounts for 50-102 Kg COD m^{-3} and 27-60 kg BOD_5 m^{-3} with a BOD_5/COD>0.5 (Prazeres et al., 2012). CWW is produced in large quantities (9 kg CW per Kg) cheese produced according to (Siso 1996) or 2.3L waste water per kg cheese produced after whey recovery according to Economopoulos. CWW management has been restricted mostly to disposal practices. However technology develops focus turns to upgrading the potential of CWW major components (lactose and proteins). Moreover, the stringent legislation on the waste disposal contributes into exploring alternatives to recover the or to produce high added value products as well as energy from CWW.

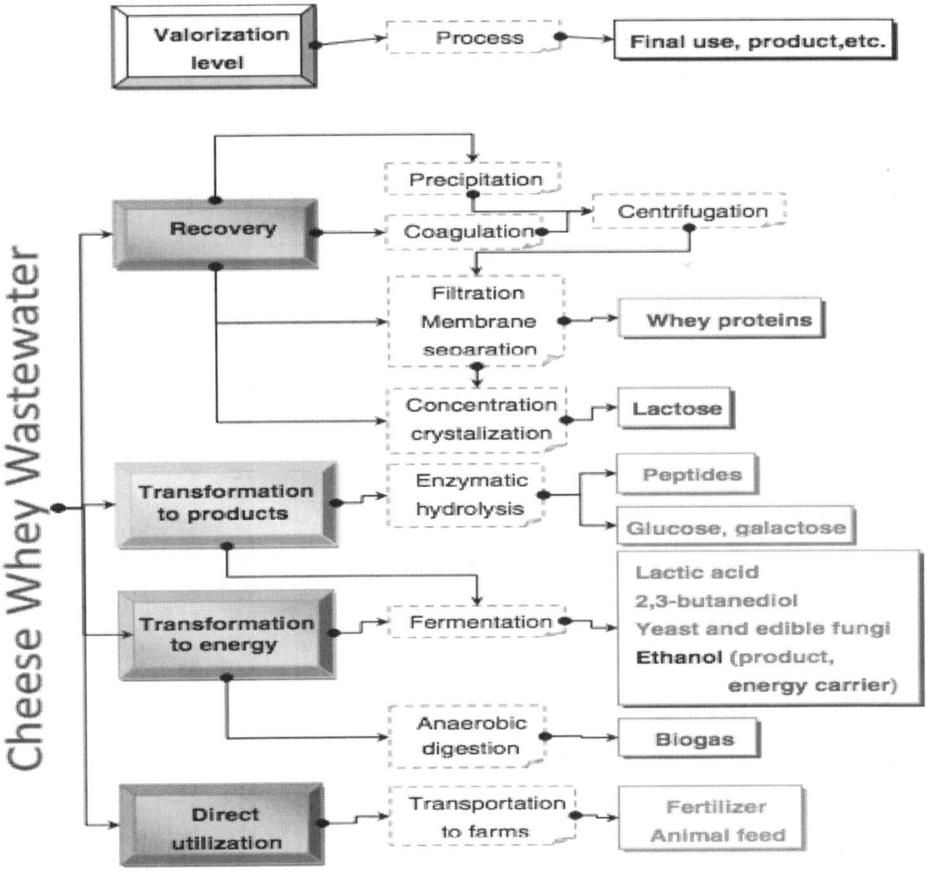

Figure 2: CWW Valorization process.

Applications And Benefits Of Encapsulated Products Derived From Food Industry Waste And By-Products

The incorporation of these components into food products is complicated by presence of bioactive substances in the food by-products, such as phenolic compounds, pro-cyanidins, polyphenols, anthocyanins, vitamins, and polyunsaturated fatty acids, which are sensitive to the environmental factors. Encapsulation has been widely utilized the method of incorporating food by-products into foods, according to the previous research (Ozkan 2019). These by-products can be successfully incorporated into food items since they are high in bioactive ingredients. The bioactive food products have been produced as a functional food which may contain an antioxidant, antibacterial, neurotransmitter, anti-diabetic, antifungal, and other qualities. These by-products are included in a variety of food items, it includes bakery goods like cookies, cakes, muffins, etc., animal products like beef, chicken, meat, sausages, etc., dairy items like cheese, yoghurt, curd, butter, ice cream, and beverages like orange, apple, and carrot juices.

Summary And Future Trends

The By-products from food sector will open up a new market for functional food additives. One of the most significant challenges in food science and technology is search for novel functional food components derived from the natural sources. By creating a valuable food by-products and boosting profitability by using different science and innovation, by- products of food industry are an excellent source of proteins, minerals, fatty acids, fibers, and bioactive compounds and substances. By-product from the food industry are significant because they can be used as a key raw material for the development of functional foods. The use of encapsulation technology is one of the current initiatives to promote the ingredient stability and the controlled release of the bioactive chemicals, and an extended product shelf-life.

Although there are very few negative side effects from taking natural supplements, authorities must exercise caution because goods are increasingly being tampered with to include bioactive ingredients. In addition to the absence of regulatory or legal guidelines for the utilization of food waste, this research compares to the regulatory environment and outline of the risk assessment standards that the bioactive chemicals are associated with food must adhere in order to maintain the product safety. The food sector now supports a culture and nation with zero waste.

BIBLIOGRAPGHY

Alabidein, Z., Ceyhan, T., & Yildirim Yalcin, M. (2024). MICROENCAPSULATION OF MANDARIN PEEL EXTRACT USING DIFFERENT METHODS.

Aswin, G., Bhasin, A., & Mazumdar, A. (2022). Utilization of jackfruit by-products and application in food industry. Pharma Innov. J, 11, 2293-2299.

DECISIVE Project. "Decentralized management of organic waste for sustainable cities." Available online: https://www.decisive2020.eu/an Environmental Agency (EEA). "The role of community-led biogas and composting in Europe's circular economy." Circular Economy Reports, 2021 .

Duysak, L., & Kılıç, A. F. (2024). Oxidative stress in chronic diseases: An Overview of Orange Peel Extracts. Current Research in Health Sciences, 1(1), 25-33.

Osorio, L. L. D. R., Flórez-López, E., & Grande-Tovar, C. D. (2021). The potential of selected agri-food loss and waste to contribute to a circular economy: Applications in the food, cosmetic and pharmaceutical industries. Molecules, 26(2), 515.

Abd El-Galil, A. A., Negm El-Dein, A., Awad, H. M., & Helmy, W. A. (2021). Chemical composition and biological activities of aqueous extracts and their sulfated derivatives of pea peel (Pisum sativum L.). Biocatalysis and Agricultural Biotechnology, 35, 102077. https://doi.org/10.1016/j.bcab.2021.102077

Abdurrahman, M. I., Chaki, S., & Saini, G. (2020). Stubble burning: Effects on health & environment, regulations and management practices. Environmental Advances, 2, 100011.

Abrol, G. S., Singh, A. K., Pal, R., Kumar, A., Sharma, P., & Sharma, G. (2023). Utilization of bottle gourd (Lagenaria siceraria (Mol.) Standl.) pomace for the preparation of instant kheer (dessert) mix. Heliyon, 9(8), e18533. https://doi.org/10.1016/j.heliyon.2023.e18533

Acharya, S., Jakeer, S., Shilpa, P., & Andhale, S. (2021). Review: Flavor encapsulation by spray drying technique. International Journal of Chemical Studies, 9(1), 1836–1840. https://doi.org/10.22271/chemi.2021.v9.i1z.11492

Agamuthu, P. (2009, November). Challenges and opportunities in agro-waste management: An Asian perspective. In Inaugural meeting of first regional 3R forum in Asia (Vol. 11, p. 12). University of Malasiya.

Agamuthu, P., Mehran, S. B., Norkhairah, A., & Norkhairiyah, A. (2019). Marine debris: A review of impacts and global initiatives. Waste Management and Research, 37(10): 987-1002.

Aguilera JF. 1989. Use of agro-industrial by-products in the feeding of ruminants. Revista Argentina de production animal 9: 253–267.

Ahamed, F., Riaz, M., Malik, M. A., & Rehman, S. (2016). Recycling and reuse of agricultural waste in soil health. Journal of Environmental Management, 182, 294–308.

Ahmed, Z., El-Sharnouby, G., & EL-Waseif, M. (2021). Use of Banana Peel As A By-Product to Increase The Nutritive Value of The Cake. Journal of Food and Dairy Sciences, 12(4), 87–97. https://doi.org/10.21608/jfds.2021.167053

Ajila, C. M., Brar, S. K., Verma, M., Tyagi, R. D., Godbout, S., & Valéro, J. R. (2012). Bio-processing of agro-byproducts to animal feed. Critical Reviews In Biotechnology, 32(4): 382-400.

Aka, S., Buyukdag, N., 2021. How to prevent food waste behaviour? A deep empirical research. J. Retailing Consum. Serv. 61, 102560.

Akdeniz, B., Sumnu, G., & Sahin, S. (2018). Microencapsulation of phenolic compounds extracted from onion (Allium cepa) skin. Journal of food Processing and Preservation, 42(7), e13648.

Alibekov, R. S., Mustapa Kamal, S. M., Taip, F. S., Sulaiman, A., Azimov, A. M., & Urazbayeva, K. A. (2023). Recovery of Phenolic Compounds from Jackfruit Seeds Using Subcritical Water Extraction. Foods, 12(17), 3296.

Amoah, J. O., Britwum, A. O., Essaw, D. W., & Mensah, J. (2023). Solid waste management and gender dynamics: Evidence from rural Ghana. Research in Globalization, 6, 100111.

Amulya, P. R., & ul Islam, R. (2023). Optimization of enzyme-assistedextraction of anthocyanins from eggplant (Solanum melongena L.) peel. Food Chemistry: X, 18, 100643.

Anastopoulos, I., Bhatnagar, A., Lima, E. C., & Vasilakos, S. (2019). Valorization of agricultural waste and by-products for the production of bio-based products. Waste Management & Research, 37(11), 982–993.

Andrade, R. A. M. de S., da Silva, D. C., Souza, M. M. B. de, de Oliveira, R. L., Maciel, M. I. S., Porto, A. L. F., Melo, E. de A., Arruda, L. L. de A. L., & Porto, T. S. (2023). Microencapsulation of phenolic compounds from cashew apple (Anacardium occidentale L.) agro-food waste: Physicochemical characterization, antioxidant activity, biodisponibility and stability. Food Chemistry Advances, 3, 100364. https://doi.org/10.1016/j.focha.2023.100364

Andrade, R. C., Ribeiro, C. D. F., de Souza Caetano, V. C., Fernandes, S. S., & Otero, D. M. (2024). Trends in Dragon Fruit Peel Compound Extraction and Technological Applications. Trends in Food Science & Technology, 104721.

Appels, L., Baeyens, J., Dewil, R., & Degrève, J. (2011). Principles and potential applications of anaerobic digestion of waste-activated sludge. Progress in Energy and Combustion Science, 37(4), 429-451.

Arancon, R. A., Lin, C. S. K., Chan, K. M., & Luque, R. (2013). From waste to resource: Biorefineries and agricultural waste valorization. Biotechnology Advances, 31(6), 1413–1424.

Auras, R., Harte, B., Selke, S. E. M., & Singh, S. P. (2011). Biobased polymers: synthesis and biomedical applications. Progress in Polymer Science, 36(7), 1163-1243.

Ayala-Zavala, J.F.; Vega-Vega, V.; Rosas-Domínguez, C.; Palafox-Carlos, H.; Villa-Rodriguez, J.A.; Siddiqui, M.W.; Dávila-Aviña, J.E.; González-Aguilar, G.A. Agro-industrial potential of exotic fruit byproducts as a source of food additives. Food Res. Int. 2011, 44, 1866–1874.

Aziz, A., Noreen, S., Khalid, W., Ejaz, A., Faiz ul Rasool, I., Maham, Munir, A., Farwa, Javed, M., Ercisli, S., Okcu, Z., Marc, R. A., Nayik, G. A., Ramniwas, S., & Uddin, J. (2023). Pumpkin and Pumpkin Byproducts: Phytochemical Constitutes, Food Application and Health Benefits. ACS Omega, 8(26), 23346–23357. https://doi.org/10.1021/acsomega.3c02176

Azmat, F., Imran, A., Islam, F., Afzaal, M., Zahoor, T., Akram, R., ... & Asif Shah, M. (2023). Valorization of the phytochemical profile, nutritional composition, and therapeutic potentials of garlic peel: a concurrent review. International Journal of Food Properties, 26(1), 2642-2655.

Baez-Rogelio, A., Morales-García, Y. E., Quintero-Hernández, V., & López-Bucio, J. (2016). Microbial valorization of agricultural residues and waste for bioproducts. Bioresource Technology, 204, 89–97.

Bakshi, M. P. S., Wadhwa, M., & Makkar, H. P. (2016). Waste to worth: vegetable wastes as animal feed. CABI Reviews, (2016), 1-26.

Banerjee, A., Rai, P. K., & Singh, O. V. (2017). Microbial degradation of plastics: A comprehensive review. 3 Biotech, 7(1), 1-19.

Bano i , M., Krzywonos, M., Aladi , K., Pi kowska, H., Mucha, I., Złoci ska, A., & Joki , S. (2023). Physicochemical, structural characterization and evaluation of encapsulated hesperidin from natural sources: Comparison of two encapsulation techniques; spray drying and freeze drying. Journal of Drug Delivery Science and Technology, 90, 105098. https://doi.org/10.1016/j.jddst.2023.105098

Barretto, F. J. de F. P., Clemente, H. A., Santana, A. L. B. D., & Vasconcelo, M. A. da S. (2020). Stability of encapsulated and non-encapsulated anthocyanin in yogurt produced with natural dye obtained from Solanum melongena L. Bark. Revista Brasileira de Fruticultura, 42(3). https://doi.org/10.1590/0100-29452020137

Bastos, Y., Rocha, F., & Estevinho, B. N. (2024). Microencapsulation of Extracts of Strawberry (Fragaria vesca) By-Products by Spray-Drying Using Individual and Binary/Ternary Blends of Biopolymers. Molecules, 29(19), 4528.

Basu, S., Saha, A., & Mukherjee, A. (2013). Traditional Weaving Techniques in India: A Study of Natural Fibers and Rural Crafting Practices. Indian Journal of Traditional Knowledge, 12(4), 512–517.

Bedrní ek, J., Laknerová, I., Linhartová, Z., Kadlec, J., Samková, E., Bárta, J., Bártová, V., Mráz, J., Pešek, M., Winterová, R., Vrchotová, N., T íska, J., & Smetana, P. (2019). Onion waste as a rich source of antioxidants for meat products. Czech Journal of Food Sciences, 37(4), 268–275. https://doi.org/10.17221/68/2018-CJFS

Berenguer, C. V., Andrade, C., Pereira, J. A., Perestrelo, R., & Câmara, J. S. (2022). Current challenges in the sustainable valorization of agri-food wastes: a review. Processes, 11(1), 20.

Bhandari, B. (2019). Crop residue as animal feed. Paper Review, 1-18.

Bhatnagar, A., Kumar, S., Sharma, P., & Gupta, A. (2015). Agricultural waste as a source of lignocellulosic enzymes and bioenergy production. Journal of Environmental Science and Technology, 9, 45–55.

Bhattacharjee, C., Dutta, S., & Saxena, V. K. (2020). A review on biosorptive removal of dyes and heavy metals from wastewater using watermelon rind as biosorbent. Environmental Advances, 2, 100007.

Bisht, B., Gururani, P., Aman, J., Vlaskin, M. S., Anna I, K., Irina A, A., Joshi, S., Kumar, S., & Kumar, V. (2023). A review on holistic approaches for fruits and vegetables biowastes valorization. Materials Today: Proceedings, 73, 54–63. https://doi.org/10.1016/j.matpr.2022.09.168

Blanco-Canqui, H., Lal, R., Owens, L. B., & Post, W. M. (2004). Soil health and environmental benefits of biochar application. Soil Biology & Biochemistry, 36(9), 2350–2357.

Bokaria, K., & Ray, S. (2016). Development of papaya peel flour based cookies and evaluation of its quality. Development, 3(12), 6393-6396.

Borah, M. S., Tiwari, A., Sridhar, K., Narsaiah, K., Nayak, P. K., & Stephen Inbaraj, B. (2023). Recent trends in valorization of food industry waste and by-products: Encapsulation and in vitro release of bioactive compounds. Foods, 12(20), 3823.

Boudjabi, S., Chehat, F., Belhamel, M., & Rahmoune, M. (2021). Environmental life cycle assessment of agricultural waste valorization pathways. Journal of Cleaner Production, 292, 126057.

Bourbon, A. I., Cerqueira, M. A., & Vicente, A. A. (2016). Encapsulation and controlled release of bioactive compounds in lactoferrin-glycomacropeptide nanohydrogels: Curcumin and caffeine as model compounds. Journal of Food Engineering, 180, 110–119. https://doi.org/10.1016/j.jfoodeng.2016.02.016

Brahma, R., & Ray, S. (2022). In-depth analysis on potential applications of jackfruit peel waste: a systematic approach. Food Chemistry Advances, 1, 100119.

Brandt, A. R., Jensen, P. A., Dam-Johansen, K., & Hustad, J. E. (2009). Gasification of biomass: Fundamentals and current status. Biomass and Bioenergy, 33(4), 316-332.

Brian L, Craig H, James L, Lisa K, Richard W, Tim S (2013) Reducing food loss and waste - installment 2 of "Creating a Sustainable Food Future." World Resource Institute, Washington DC.

Briggs, H., Lindeberg, J.D., Rein, A., Chorn, B., Tanger, K., 2016. The ReFED Roadmap to Reducing Food Waste. Available at: https://refed.com/articles/the-refed-roadma p-to-reducing-food-waste/ (Accessed on February 19, 2023).

Brown and Root Environmental Consultancy Group. (1997). Environmental Review of National Solid Waste Management Plan. Interim Report Submitted to the Government of Mauritius.

Brown, T., Smith, R., & Jones, L. (2020). Circular Bioeconomy Alliance: A global framework for waste valorization. Journal of Sustainable Development, 18(4), 89–102. https://doi.org/10.xxxx

Buljeta, I., Pichler, A., Šimunovi , J., & Kopjar, M. (2022). Polysaccharides as Carriers of Polyphenols: Comparison of Freeze-Drying and Spray-Drying as Encapsulation Techniques. Molecules, 27(16), 5069. https://doi.org/10.3390/molecules27165069

Byrne, E., Curran, T., Williams, I., & McKeown, L. (2020). Circular economy practices in agricultural production systems. Resources, Conservation and Recycling, 155, 104716.

C. Chen, A. Chaudhary and A. Mathys, Nutritional and environmental losses embedded in global food waste, Resour., Conserv. Recycl., 2020, 160, 104912, DOI: 10.1016/ j.resconrec.2020.104912.

Capanoglu, E., Nemli, E., & Tomas-Barberan, F. (2022). Novel approaches in the valorization of agricultural wastes and their applications. Journal of Agricultural and Food Chemistry, 70(23), 6787–6804.

Capanoglu, E.; Nemli, E.; Tomas-Barberan, F. Novel Approaches in the Valorization of Agricultural Wastes and Their Applications. J. Agric. Food Chem. 2022, 70, 6787–6804.

Caponio, G. R., Minervini, F., Tamma, G., Gambacorta, G., & De Angelis, M. (2023). Promising application of grape pomace and its agri-food valorization: sourceof bioactive molecules with beneficial effects. Sustainability, 15(11), 9075.

Caputo, V., De Pascale, D., & Grassi, M. (2019). Smart waste management: A review. Waste Management, 94, 510-521.

Carreón-Delgado, D. F., Hernández-Montesinos, I. Y., Rivera-Hernández, K. N., del Sugeyrol Villa-Ramírez, M., Ochoa-Velasco, C. E., & Ramírez-López, C. (2023). Evaluation of pretreatments and extraction conditions on the antifungal and antioxidant effects of garlic (Allium sativum) peel extracts. Plants, 12(1), 217.

Carrera, C., Pastol, J., Setyaningsih, W., Ruiz-Rodríguez, A., Ferreiro-González, M., Fernández Barbero, G., & Palma, M. (2021). Optimization by means of chemometric tools of an ultrasound-assisted method for the extraction of betacyanins from red dragon fruit (Hylocereus polyrhizus). Agronomy, 11(6), 1053.

Cassani, L., Marcovich, N. E., & Gomez-Zavaglia, A. (2022). Valorization of fruit and vegetables agro-wastes for the sustainable production of carotenoid-based colorants with enhanced bioavailability. Food Research International, 152, 110924. https://doi.org/10.1016/j.foodres.2021.110924

Cazón, P., & Sanches Silva, A. (2024). Natural pigments from food wastes: New approaches for the extraction and encapsulation. Current Opinion in Green and Sustainable Chemistry, 47, 100929. https://doi.org/10.1016/j.cogsc.2024.100929

Cerda, A., Gómez, J. A., Morera, A., & Navarro, R. (2017). Water pollution prevention in waste valorization systems. Water Research, 126, 72–85.

Chaari, M., Elhadef, K., Akermi, S., Hlima, H. B., Fourati, M., Chakchouk Mtibaa, A., ... & Smaoui, S. (2023). Multiobjective response and chemometric approaches to enhance the phytochemicals and biological activities of beetroot leaves: an unexploited organic waste. Biomass Conversion and Biorefinery, 13(16), 15067-15081.

Chaudhary, B. U., Lingayat, S., Banerjee, A. N., & Kale, R. D. (2021). Development of multifunctional food packaging films based on waste Garlic peel extract and Chitosan. International Journal of Biological Macromolecules, 192, 479–490. https://doi.org/10.1016/j.ijbiomac.2021.10.031

Chauhan, A., Upadhyay, S., Saini, G., & Senthilkumar, N. (2022). Agricultural crop residue based biomass in India: Potential assessment, methodology and key issues. Sustainable Energy Technologies and Assessments, 53, 102552.

Chemat, F., et al. "Green extraction of bioactive compounds from food by-products." Innovative Food Science & Emerging Technologies, 41, 56-67, 2017

Chemat, F., Vian, M. A., & Camara, L. S. (2012). Green extraction of natural products: Membrane technologies. Trends in Analytical Chemistry, 31(1), 3-17.

Chen, Y., Wen, J., Deng, Z., Pan, X., Xie, X., & Peng, C. (2020). Effective utilization of food wastes: Bioactivity of grape seed extraction and its application in food industry. Journal of Functional Foods, 73, 104113.

Chenost, M., & Sansoucy, R. (1998). Nutritional characteristics of tropical feed resources. Natural and improved grasslands, crop residues and agro industrial byproducts. Feeding dairy cows in the tropics.

Chihoub, W., Dias, M. I., Barros, L., Calhelha, R. C., Alves, M. J., Harzallah-Skhiri, F., & Ferreira, I. C. F. R. (2019). Valorization of the green waste parts from turnip, radish and wild cardoon: Nutritional value, phenolic profile and bioactivity evaluation. Food Research International, 126, 108651. https://doi.org/10.1016/j.foodres.2019.108651

Chowdhury, F., Rashid, A., Hossain, M., & Karim, S. (2016). Assessment of agricultural waste valorization in Asia. Renewable & Sustainable Energy Reviews, 66, 865–877.

Comunian, T. A., Silva, M. P., & Souza, C. J. F. (2021). The use of food by-products as a novel for functional foods: Their use as ingredients and for the encapsulation process. Trends in Food Science & Technology, 108, 269–280. https://doi.org/10.1016/j.tifs.2021.01.003

Constantin, O. E., Milea, A. ., Bolea, C., Mihalcea, L., Enachi, E., Copolovici, D. M., ... & Râpeanu, G. (2021). Onion (Allium cepa L.) peel extracts characterization by conventional and modern methods. International Journal of Food Engineering, 17(6), 485-493.

Culas, M. S., Popovich, D. G., & Rashidinejad, A. (2024). Recent advances in encapsulation techniques for cinnamon bioactive compounds: A review on stability, effectiveness, and potential applications. Food Bioscience, 57, 103470. https://doi.org/10.1016/j.fbio.2023.103470

Dai, Y., Sun, Q., Wang, W., Lu, L., Liu, M., Li, J., and Zhang, Y. (2018). Utilizations of agricultural waste as adsorbent for the removal of contaminants: A review. Chemosphere, 211, 235-253.

Dale, B.E., et al. "The potential for expanding sustainable biogas production." Biofuels, Bioprod. Bioref. 14, 1335–1347, 2020.

Dantas, T. L., Alonso Buriti, F. C., & Florentino, E. R. (2021). Okra (Abelmoschus esculentus L.) as a Potential Functional Food Source of Mucilage and Bioactive Compounds with Technological Applications and Health Benefits. Plants,10(8), 1683. https://doi.org/10.3390/plants10081683

Dastjerdi, F., Mousavi, S., Taghavi, H., & Khademian, A. (2021). Agricultural waste as a feedstock for the production of biofuels. Renewable & Sustainable Energy Reviews, 135, 110276.

de Sena Andrade, R. A. M., da Silva, D. C., de Souza, M. M. B., de Oliveira, R. L., Maciel, M. I. S., Porto, A. L. F., ... & Porto, T. S. (2023). Microencapsulation of phenolic compounds from cashew apple (Anacardium occidentale L.) agro-food waste: Physicochemical characterization, antioxidant activity, biodisponibility and stability. Food Chemistry Advances, 3, 100364.

Decorato, D., Braguglia, C. M., Luciani, G., & Mininni, G. (2020). Reducing emissions from agricultural waste valorization. Environmental Science & Technology, 54(2), 987–995.

Demirba A. Waste management, waste resource facilities and waste conversion processes. Energy Conversion and Management. 2011; 52:1280-1287. DOI-10.1016/j.enconman.2010.09.025.

Demirbas, A. (2011). Biochar as a Sustainable Energy Source. Renewable and Sustainable Energy Reviews, 15(9), 4444–4450.

Desmae, H., and Sones, H. (2017). Groundnut cropping guide.

Dignac, M., Gaillard, V., Rumpel, C., & Marschner, B. (2017). Soil organic matter and the role of biochar in enhancing soil properties. Agricultural Systems, 156, 12–21.

Dobermann, A. T. H. F., and Fairhurst, T. H. (2002). Rice straw management. Better crops international, 16(1), 7-11.

dos Santos, P. C. M., da Silva, L. M. R., Magalhaes, F. E. A., Cunha, F. E. T., Ferreira, M. J. G., & de Figueiredo, E. A. T. (2022). Garlic (Allium sativum L.) peel extracts: from industrial by-product to food additive. Applied Food Research, 2(2), 100186.

Drosou, C., & Krokida, M. (2024). A Comparative Study of Encapsulation of -Carotene via Spray-Drying and Freeze-Drying Techniques Using Pullulan and Whey Protein Isolate as Wall Material. Foods, 13(12), 1933. https://doi.org/10.3390/foods13121933

Drosou, C., Kyriakopoulou, K., Bimpilas, A., Tsimogiannis, D., & Krokida, M. (2015). A comparative study on different extraction techniques to recover red grape pomace polyphenols from vinification byproducts. Industrial Crops and Products, 75, 141-149. 3. Zaini, H. M., Roslan, J., Saallah, S., Munsu, E., Sulaiman, N. S., & Pindi, W. (2022). Banana peels as a bioactive ingredient and its potential application in the food industry. Journal of Functional Foods, 92, 105054.

Duan, Y., Liu, X., Zhang, M., & Wang, H. (2021). The use of agricultural wastes as a biofertilizer source to mitigate greenhouse gas emissions. Agronomy, 11(12), 1449.

Dunbar, P. Increasing public awareness of natural hazards via the Internet. Nat. Hazards 2007, 42, 529–536.

Dung, N. T., Pham, H. V., Tran, T. M., & Nguyen, Q. T. (2020). Application of biofertilizers in circular agricultural systems. Agronomy, 10(5), 489.

Durandeau, S., Gabrielle, B., Godard, C., Jayet, P. A., & Le Bas, C. (2010). Coupling biophysical and micro-economic models to assess the effect of mitigation measures on greenhouse gas emissions from agriculture. Climatic Change, 98(1), 51–73.

Durazzo, A., Lucarini, M., Novellino, E., Souto, E. B., Daliu, P., & Santini, A. (2018). Abelmoschus esculentus (L.): Bioactive Components' Beneficial Properties—Focused on Antidiabetic Role—For Sustainable Health Applications. Molecules, 24(1), 38. https://doi.org/10.3390/molecules24010038

Echave, J., Pereira, A. G., Carpena, M., Prieto, M. Á., & Simal-Gandara, J. (2020). Capsicum seeds as a source of bioactive compounds: biological properties, extraction systems, and industrial application. In Capsicum (pp. 1-37). London: IntechOpen.

El-Beltagi, H.S.; Basit, A.; Mohamed, H.I.; Ali, I.; Ullah, S.; Kamel, E.A.R.; Shalaby, T.A.; Ramadan, K.M.A.; Alkhateeb, A.A.; Ghazzawy, H.S. Mulching as a Sustainable Water and Soil Saving Practice in Agriculture: A Review. Agronomy 2022, 12, 1881. https:// doi.org/10.3390/agronomy12081881

Elferink, E.V.; Nonhebel, S.; Moll, H.C. Feeding livestock food residue and the consequences for the environmental impact of meat. J. Clean. Prod. 2008, 16, 1227–1233.

El-Messery, T. M., Altuntas, U., Altin, G., & Özçelik, B. (2020). The effect of spray-drying and freeze-drying on encapsulation efficiency, in vitro bioaccessibility and oxidative stability of krill oil nanoemulsion system. Food Hydrocolloids, 106, 105890. https://doi.org/10.1016/j.foodhyd.2020.105890

El-Messery, T. M., El-Said, M. M., Demircan, E., & Ozçelik, B. (2019). Microencapsulation of natural polyphenolic compounds extracted from apple peel and its application in yoghurt. Acta Scientiarum Polonorum Technologia Alimentaria, 18(1), 25-34.

El-Sayed, S. M. (2020). Use of spinach powder as functional ingredient in the manufacture of UF-Soft cheese. Heliyon, 6(1).

Elsebaie, E. M., & Essa, R. Y. (2018). Microencapsulation of red onion peel polyphenols fractions by freeze drying technicality and its application in cake. Journal of Food Processing and Preservation, 42(7), e13654. https://doi.org/10.1111/jfpp.13654

Elsebaie, E. M., & Essa, R. Y. (2018). Microencapsulation of red onion peel polyphenols fractions by freeze drying technicality and its application in cake. Journal of Food Processing and Preservation, 42(7), e13654. https://doi.org/10.1111/jfpp.13654

Espinosa Alvarez, C. S., López Contreras, J., Escobar Rodríguez, D., Jiménez Rondón, D., Bugueño Muñoz, W., & Cerezal Mezquita, P. (2019). Application of microencapsulated anthocyanin extracted from purple cabbage in fermented milk drinks. Acta Agronómica, 68(2), 134–141.https://doi.org/10.15446/acag.v68n2.79078

Eun, J.-B., Maruf, A., Das, P. R., & Nam, S.-H. (2020). A review of encapsulation of carotenoids using spray drying and freeze drying. Critical Reviews in Food

Science and Nutrition, 60(21), 3547–3572. https://doi.org/10.1080/10408398.2019.1698511

European Commission. "Biogas and composting as sulutions for urban waste management in Germany." Available online: https://www.ec.europa.eu/environment .

European Commission. (2018). The Circular Economy Action Plan. Retrieved from https://eur-lex.europa.eu/legal-content/EN/TXT/?uri=CELEX:52015DC0614

European Commission. (2021). Horizon 2020 and the bioeconomy. Retrieved from https://ec.europa.eu/

European Commission. EU Platform on Food Losses and Food Waste. Available online: https://ec.europa.eu/food/safety/food- waste/eu-actions-against-food-waste/eu-platform-food-losses-and-food-waste_ro.

European Commission. Innovating for sustainable growth: A bioeconomy for Europe. COM (212) 60 final. Available online: http://ec.europa.eu/research/bioeconomy/pdf/201202_innovating_sustainable_growth.pdf (accessed on 6 July 2015).

European Union. Directive No 2008/98 of the European Parliament and the Council of 19 November 2008 on Waste and Repealing Certain Directives. Off. J. Eur. Union 2008, L312, 3–30.

Eyob, H., Njonge, F. K., Goitom, A., & Gicheha, M. (2016). Production and potential utilization of crop residues and agro-industrial byproducts in ruminant nutrition in Eritrea.

Ezzat, S. M., Adel, R., & Abdel-Sattar, E. (2022). Pumpkin bio-wastes as source of functional ingredients. In Mediterranean fruits bio-wastes: chemistry, functionality and technological applications (pp. 667-696). Cham: Springer International Publishing.

F. O. Obi1, B. O. Ugwuishiwu2 and J. N Nwakaire3, AGRICULTURAL AND BIORESOURCES ENGINEERING DEPT, UNIVERSITY OF NIGERIA, NSUKKA. ENUGU STATE. NIGERIA, Nigerian Journal of Technology (NIJOTECH) Vol. 35, No. 4, October 2016, pp. 957 – 964

Fahmy, Y., Mobarak, F., El-Sakhawy, M., and Fadl, M. H. (2017). Agricultural residues (wastes) for manufacture of paper, board, and miscellaneous products: Background overview and future prospects. International journal of chemtech research, 10(2), 424-448.

Falade, A. O. (2021). Valorization of agricultural wastes for production of biocatalysts of environmental significance: Towards a sustainable environment. Environmental Sustainability, 4(2), 317–328.

FAO (2011) Global food losses and food waste – extent, causes and prevention.

FAO (2012) Towards the future we want: end hunger and make the transition to sustainable agricultural and food systems. Food and Agriculture Organization of the United Nations Rome.

FAO, 2023. Food Wastage Footprint: Full Cost-Accounting. FAO, Rome, Italy. https://www.fao.org/documents/card/en?details=5e7c4154-2b97-4ea5-83a7-be9604925a24%2F (Accessed 2 December 2023).

FAO. Food Wastage Footprint: Impacts on Natural Resources; Summary Report; Food and Agriculture Organization of the United Nations: Rome, Italy, 2013; Available online: http://www.fao.org/3/i3347e/i3347e.pdf.

Farooq, M., Anwar, R., Shukat, R., QamarAbbas, S., Ilyas, N., Cristina, M., & Wang, Y. Y. (2021). Ecofriendly utilization of by products from banana peel in food production and other industrial applications. a review.

Farooque, S., Rose, P. M., Benohoud, M., Blackburn, R. S., & Rayner, C. M. (2018). Enhancing the potential exploitation of food waste: Extraction, purification, and characterization of renewable specialty chemicals from blackcurrants (Ribes nigrum L.). Journal of agricultural and food chemistry, 66(46), 12265-12273.

Fatima, R., Fatima, F., Altemimi, A. B., Bashir, N., Sipra, H. M., Hassan, S. A., Mujahid, W., Shehzad, A., Abdi, G., & Aadil, R. M. (2024). Bridging sustainability and industry through resourceful utilization of pea pods- A focus on diverse industrial applications. Food Chemistry: X, 23, 101518. https://doi.org/10.1016/j.fochx.2024.101518

Ferarsa, S., Zhang, W., Moulai-Mostefa, N., Ding, L., Jaffrin, M. Y., & Grimi, N. (2018). Recovery of anthocyanins and other phenolic compounds from purple eggplant peels and pulps using ultrasonic-assisted extraction. Food and Bioproducts Processing, 109, 19-28.

Fernández, R., Martínez, A., & López, J. (2019). Domestic markets for bio-based products: A growing trend. Bioeconomy Journal, 12(3), 45–62. https://doi.org/10.xxxx

Fernández-López, J.; Sendra, E.; Sayas-Barberá, E.; Navarro, C.; Pérez-Alvarez, J.A. Physico-chemical and microbiological profiles of "salchich" (Spanish dry-fermented sausage) enriched with orange fiber. Meat Sci. 2008, 80, 410–417.

Fertahi, S., Elalami, D., Tayibi, S., Taarji, N., Lyamlouli, K., Bargaz, A., and Barakat, A. (2023). The current status and challenges of biomass biorefineries in Africa: A critical review and future perspectives for bioeconomy development. Science of The Total Environment, 870, 162001.

Fiorentini, C., Duserm Garrido, G., Bassani, A., Cortimiglia, C., Zaccone, M., Montalbano, L., Martinez-Nogues, V., Cocconcelli, P. S., & Spigno, G. (2021). Citrus Peel Extracts for Industrial-Scale Production of Bio-Based Active Food Packaging. Foods, 11(1), 30. https://doi.org/10.3390/foods11010030

Fontana, A. R., Antoniolli, A., & Bottini, R. (2013). Grape pomace as a sustainable source of bioactive compounds: Extraction, characterization, and bitechnological applications of phenolics. Journal of agricultural and food chemistry, 61(38), 8987-9003.

Food and Agriculture Organization (FAO), 2021. Save Food: Global Initiative on Food Loss and Waste Reduction. FAO, Rome, Italy. https://www.fao.org/save-food/en/ (Accessed 19 December 2023).

Food and Agriculture Organization (FAO). World Food Security: a Reappraisal of the Concepts and Approaches. Direct General's Report. Rome; 1983.

Food and Nutrition Technical Assistance. Available online: http://www.fantaproject.org (accessed on 15 October 2012).

Francesco Montemurro1 and Angelo Fiore., Gabriele Campanelli., Fabio Tittarelli., Luigi Ledda., Stefano Canali(2013). Organic Fertilization, Green Manure, and Vetch Mulch to Improve Organic Zucchini Yield and Quality. HORTSCIENCE 48(8):1027–1033. 2013

Frankenberger T, Frankel L, Ross S, Marshall B, Cardenas C, Clark D, et al. Household livelihood security. a unifying conceptual framework for CARE programs. In: Proceedings of the USAID Workshop on Performance Measurement for Food Security. 11-12 December; 1995.

Fredes, C., Becerra, C., Parada, J., & Robert, P. (2018). The Microencapsulation of Maqui (Aristotelia chilensis (Mol.) Stuntz) Juice by Spray-Drying and Freeze-Drying Produces Powders with Similar Anthocyanin Stability and Bioaccessibility. Molecules, 23(5), 1227. https://doi.org/10.3390/molecules23051227

Fuentes, Y., Giovagnoli-Vicuña, C., Faúndez, M., & Giordano, A. (2023). Microencapsulation of Chilean Papaya Waste Extract and Its Impact onPhysicochemical and Bioactive Properties. Antioxidants, 12(10), 1900. https://doi.org/10.3390/antiox12101900

Galanakis, C.M. Recovery of high added-value components from food wastes: Conventional, emerging technologies and commercialized applications. Trends Food Sci. Technol. 2012, 26, 68–87.

Garcia, M., Brown, H., & Wilson, T. (2018). AgroCycle: Waste to wealth in European agriculture. Environmental Innovation and Societal Transitions, 27, 1–15. https://doi.org/10.xxxx

García, S. L. R. (2022). Microwave-Assisted Extraction of Phenolic Compounds from Broccoli «Brassica Oleracea» By-Products (Leaves and Stems), and Broccoli Edibles (Florets). McGill University (Canada).

Gautam, M.; Agrawal, M. Greenhouse Gas Emissions from Municipal Solid Waste Management: A Review of Global Scenario. In Carbon Footprint Case Studies: Environmental Footprints and Eco-Design of Products and Processes; Muthu, S., Ed.; Springer Nature: Singapore, 2021.

Gavril, R. N., Stoica, F., Lip a, F. D., Constantin, O. E., St nciuc, N., Aprodu, I., & Râpeanu, G. (2024). Pumpkin and Pumpkin By-Products: AComprehensive Overview of Phytochemicals, Extraction, Health Benefits, and Food Applications. Foods, 13(17), 2694.

German Federal Egency. "Advancements in composting and biogas production at the community level." Environmental Innovation Reports, 2019 .

Girotto, F., & Cossu, R. (2017). Animal waste: Opportunities and challenges. Sustainable agriculture reviews, 1-13.

Gonzales, C. M. (2024). Microencapsulation of bioactive compounds from avocado oil residues (Doctoral dissertation, Universidade de SÃ£ o Paulo).

Government of India. (2021). National Biogas and Manure Management Program guidelines. Retrieved from https://www.mnre.gov.in/

Goyal, S., Kumar, S., & Mittal, A. (2017). Smart waste management: A review. Waste Management, 69, 11-20.

Goyeneche, R., Di Scala, K., Ramirez, C. L., & Fanovich, M. A. (2020). Recovery of bioactive compounds from beetroot leaves by supercritical CO2 extraction as a promising bioresource. The Journal of Supercritical Fluids, 155, 104658.

Grassino, A. N., Pedisi , S., Dragovi -Uzelac, V., Karlovi , S., Je ek, D., & Bosiljkov, T. (2020). Insight into high-hydrostatic pressure extraction of polyphenols from tomato peel waste. Plant foods for human nutrition, 75, 427-433.

Grewal, L., Hmurovic, J., Lamberton, C., Reczek, R.W., 2019. The self-perception connection. J. Market. 83 (1), 89–107.

Grover P, Singh P (2014) An analytical study of effect of family income and size on per capita household solid waste generation in devel- oping countries. Rev Arts Human 3(1):127–143.

Guebebia, S., Gharsallaoui, A., Dumas, E., Baghi, F., Zourgui, L., Romdhane, M., ... & Ghnimi, S. (2023). Microencapsulation of phenolic compounds extracted from okra (Abelmoschus esculentus L.) leaves, fruits and seeds. Applied Sciences, 13(22), 12273.

Guillard, V., Gaucel, S., Fornaciari, C., Angellier-Coussy, H., Buche, P., and Gontard, N. (2018). The next generation of sustainable food packaging to preserve our environment in a circular economy context. Frontiers In Nutrition, 5, 121.

Gupta, J., Kumari, M., Mishra, A., Akram, M., and Thakur, I. S. (2022). Agro-forestry waste management-A review. Chemosphere, 287, 132321.

Gupta, R., & Agrawal, A. (2014). Rice Husk Biomass: An Emerging Renewable Energy Source for Rural India. Energy for Sustainable Development, 21(4), 97-103.

Gupta, V. (2011). Sustainable Craft Practices: Use of Agricultural By-products in Indian Traditional Handicrafts. International Journal of Sustainable Development, 14(2), 65-72.

Gustavsson, J.; Cederberg, C.; Sonesson, U. Global Food Losses and Food Waste: Extent, Causes and Prevention; Food and Agriculture Organization of the United Nations; FAO: Rome, Italy, 2011; Available online: http://www.fao.org/3/i2697e/i2697e.pdf.

Hahn T, McDermott C, Ituarte-Lima C, Schultz M, Green T, Tuvendal M (2015) Purposes and degrees of commodification: economic instruments for biodiversity and ecosystem services need not rely on markets or monetary valuation. Ecosyst Serv 16:74–82. https://doi.org/10.1016/j.ecoser.2015.10.012

Haque, A. B., Islam, A. N., and Mikalef, P. (2023). Explainable Artificial Intelligence (XAI) from a user perspective: A synthesis of prior literature and problematizing avenues for future research. Technological Forecasting and Social Change, 186, 122120.

Hariadi, H., Nissa, R. C., Anggara, C. E. W., Hidayat, Ferdiansyah, A. E., & Rifqi, M. (2023). Utilization of dragon fruit (Hylocereus polyrhizus) peels waste as a natural dye. IOP Conference Series: Earth and Environmental Science, 1241(1), 012075. https://doi.org/10.1088/1755-1315/1241/1/012075

Harshwardhan K, Upadhyay K. Effective Utilization of Agricultural Waste: Review. Journal of Fundamentals of Renewable Energy and Applications, 2008, 7(5). DOI:10.4172/2090-4541.1000237

Harshwardhan, K., and Upadhyay, K. J. J. F. R. E. A. (2017). Effective utilization of agricultural waste: review. J. Fundam. Renew. Energy Application, 7(5).

Hernández Montesinos, I. Y., Carreón Delgado, D. F., Ocaranza Sánchez, E., Ochoa Velasco, C. E., Suárez Jacobo, Á., & Ramírez López, C. (2023). Garlic (Allium sativum) peel extracts and their potential as antioxidant and antimicrobial agents for food applications: influence of pretreatment and extraction solvent. International Journal of Food Science & Technology, 58(12), 6794–6805. https://doi.org/10.1111/ijfs.16652

Hinostroza-Quiñonez, G., Alcántara-Mucha, L., Espinoza-Silva, C., Quispe-Solano, M., & Saavedra-Mallma, N. (2024). Physicochemical and functional properties of microcapsules of papaya seed extract (Carica papaya L.) obtained by spray drying. NFS Journal, 35, 100164.

Hitlamani, V., & Inamdar, A. A. (2024). Valorization and the potential use of garlic (Allium sativum L.) skin in food industries. Food and Humanity, 3, 100437. https://doi.org/10.1016/j.foohum.2024.100437

Hofstede, G. Culture's Consequences: Comparing Values, Behaviors, Institutions, and Organizations Across Nations; Sage Publication: Thousand Oaks, CA, USA, 2001.

Huang, X., Liu, Y., & Wang, P. (2020). Emerging markets for waste valorization. Resources, Conservation, and Recycling, 157, 104749. https://doi.org/10.xxxx

Hussain, S., Jõudu, I., & Bhat, R. (2020). Dietary Fiber from Underutilized Plant Resources—A Positive Approach for Valorization of Fruit and Vegetable Wastes. Sustainability, 12(13), 5401. https://doi.org/10.3390/su12135401

Ibeawuchi, I. I., Iwuanyanwu, U. P., Nze, E. O., Olejeme, O. C., & Ihejirika, G. O. (2015). Mulches and organic manures as renewable energy sources for sustainable farming. Journal of Natural Sciences Research, 5(2), 139-147.

Ibrahim, R. M., Fayez, S., Eltanany, B. M., Abu-Elghait, M., El-Demerdash, A., Badawy, M. S. E. M., Pont, L., Benavente, F., & Saber, F. R. (2024). Agro-byproduct valorization of radish and turnip leaves and roots as new sources of antibacterial and antivirulence agents through metabolomics and molecular networking. Scientia Horticulturae, 328, 112924. https://doi.org/10.1016/j.scienta.2024.112924

Inoue, S., Shimoda, M., & Koyama, S. (2016). Utilization of Agricultural Residues for Paper Production: A Sustainable Approach to the Use of Natural Resources. Journal of Environmental Management, 183, 149–156.

Irina, I., Cédric, P., Ghoul, M., & Boudhrioua, N. (2017). Antioxidants of Maltease orange peel: Comparative investigation of the efficiency of four extraction methods. Journal of Applied Pharmaceutical Science, 7(11), 126-135.

Isik, B. S., Bayram, S., & Altay, F. (2024). Electrospinning Encapsulation of Chlorophylls Microwave-Extracted from Spinach Waste and Utilization as Colouring Agent in Yoghurt.

Itoiz, E. S., Gasol, C. M., Farreny, R., Rieradevall, J., & Gabarrell, X. (2013). CO2ZW: Carbon footprint tool for municipal solid waste management for policy options in Europe. Inventory of Mediterranean countries. Energy Policy, 56, 623–632.

Jaime, L., Vázquez, E., Fornari, T., López Hazas, M. D. C., García Risco, M. R., Santoyo, S., & Reglero, G. (2015). Extraction of functional ingredients from spinach (Spinacia oleracea L.) using liquid solvent and supercritical CO2 extraction.Journal of the Science of Food and Agriculture, 95(4), 722-729.

Jat, M. L., et al. (2014). „Effects of crop residue burning on the environment and health in South Asia." Agricultural Systems, 130, 25-39.

Jiang, Q., Zhao, L., & Chen, H. (2021). Economic benefits of agricultural waste valorization: A case study. Renewable Energy, 171, 23–35. https://doi.org/10.xxxx

Jiménez Bolaño, D. C., Insuasty, D., Rodríguez Macías, J. D., & Grande-Tovar, C. D. (2024). Potential Use of Tomato Peel, a Rich Source of Lycopene, for Cancer Treatment. Molecules, 29(13), 3079.

Jimenez-Champi, D., Romero-Orejon, F. L., Moran-Reyes, A., Muñoz, A. M., & Ramos-Escudero, F. (2023). Bioactive compounds in potato peels, extraction methods, and their applications in the food industry: a review. Cyta-journal of Food, 21(1), 418432.

Jiménez-Moreno, N., Esparza, I., Bimbela, F., Gandía, L. M., & Ancín-Azpilicueta, C. (2020). Valorization of selected fruit and vegetable wastes as bioactive compounds: Opportunities and challenges. Critical Reviews in Environmental Science and Technology, 50(20), 2061–2108. https://doi.org/10.1080/10643389.2019.1694819

Ju, X., Zhang, F., Bao, X., Römheld, V., & Roelcke, M. (2005). Utilization and management of organic wastes in Chinese agriculture: Past, present and perspectives. Science in China Series C: Life Sciences, 48, 965-979.

Kabeer, R. (2017). Jute Fiber and Its Use in Traditional Textiles in Southeast Asia. Journal of Cultural Heritage and Textile Studies, 3(1), 24-29.

Kainat, S., Arshad, M. S., Khalid, W., Zubair Khalid, M., Koraqi, H., Afzal, M. F., ... & Al-Farga, A. (2022). Sustainable novel extraction of bioactive compounds from fruits and vegetables waste for functional foods: a review. International Journal of Food Properties, 25(1), 2457-2476.

Kakkar, S. H. A. M. A., Tandon, R. U. N. J. H. U. N., & Tandon, N. I. T. I. N. (2021). Utilizing fruits and vegetables waste as functional food: A review. Plant Cell. Biotechnol. Mol. Biol, 22, 41-58.

Kamath, A. A., & R., N. (2018). Removal of malachite green dye in aqueous solution by Activated Custard apple (Annona Squamosa) Peel Powder. International Journal of Applied Chemistry, 14(4), 311. https://doi.org/10.37622/IJAC/14.4.2018.311-324

Kamm, B., Gruber, P. R., & Kamm, M. (2006). Biorefineries—Industrial processes and technologies for biobased products. Wiley-VCH.

Kandylis, P., & Kokkinomagoulos, E. (2020). Food applications and potential health benefits of pomegranate and its derivatives. Foods, 9(2), 122.

Karimi, A., Kazemi, M., Samani, S. A., & Simal-Gandara, J. (2021). Bioactive compounds from by-products of eggplant: Functional properties, potential applications and advances in valorization methods. Trends in Food Science & Technology, 112, 518-531.

Kaur, G. J., Kumar, D., Orsat, V., & Singh, A. (2020). Assessment of carrot rejects and wastes for food product development and as a biofuel. Biomass conversion and biorefinery, 1-12.

Kauser, S., Murtaza, M. A., Hussain, A., Imran, M., Kabir, K., Najam, A., ... & Yaqub, S. (2023). Apple pomace, a bioresource of functional and nutritional components with potential of utilization in different food formulations; a review. Food Chemistry Advances, 100598.

Kavitha, S.; Kannah, R. Y; Kumar, G.; Gunasekaran, M.; Banu, J. R. Introduction: sources and characterization of food waste and food industry wastes. In Food Waste to Valuable Resources, Banu, J. R., Kumar, G., Gunasekaran, M., Kavitha, S., Eds.; Academic Press: New York, 2020; pp 1 13.

Khaleel, R., Reddy, K. R., & Overcash, M. R. (1981). Changes in soil physical properties due to organic waste applications: a review. Journal of Environmental Quality, 10(2), 133-141.

Khan, A., Khan, S. H., Saif, M., Batool, A., Sohail, A., and Waleed Khan, M. (2023). A Survey of Deep Learning Techniques for the Analysis of COVID-19 and their usability for Detecting Omicron. Journal of Experimental and Theoretical Artificial Intelligence, 1-43

Kim, S., Dale, B. E., and Jenkins, R. (2009). Life cycle assessment of corn grain and corn stover in the United States. The International Journal of Life Cycle Assessment, 14, 160-174.

Kobo, G. K., Kaseke, T., & Fawole, O. A. (2022). Micro-encapsulation of phytochemicals in passion fruit peel waste generated on an organic farm: effect of carriers on the quality of encapsulated powders and potential for value-addition. Antioxidants, 11(8), 1579.

Koç, G. Ç., Erbakan, T., Arıcı, E., & Dirim, S. N. (2019). Sensory and quality attributes of cake supplemented with spinach powder. Gıda, 44(5), 907-918.

Kondamudi, N., Mohapatra, S. K., & Misra, M. (2008). Biodiesel production from coffee grounds. Journal of Agricultural and Food Chemistry, 56(24), 11757–11760.

Kosmala M, Kołodziejczyk K, Zdunczyk Z, Juskiewicz J, Boros D (2011) Chemical composition of natural and polyphenol-free apple pomace and the effect of this dietary ingredient on intes- tinal fermentation and serum lipid parameters in rats. Jour Agri Food Chem 59(17):9177–9185.

Kothari, S. (2015). Traditional Crafting with Agricultural Waste: A Study of Weaving Techniques in Rural India. Textile Research Journal, 17(2), 139-145.

Koul, B., Yakoob, M., & Shah, M. P. (2022). Agricultural waste management strategies for environmental sustainability. Environmental Research, 206, 112285.

Kruszewski, B., & Boselli, E. (2024). Blackcurrant Pomace as a Rich Source of Anthocyanins: Ultrasound-Assisted Extraction under Different Parameters. Applied Sciences, 14(2), 821.

KrWG cular Economy Act. "Separate collection of biowaste and waste hierarchy principles." German Federal Ministry for the Environment, Nature Conservation, and Nuclear Safety, 2015 .

Kumar, A., Gupta, R., Singh, P., & Sharma, M. (2010). Production of activated carbon from coconut shell. Journal of Environmental Management, 91(1), 204–209.

Kumar, M., Barbhai, M. D., Hasan, M., Dhumal, S., Singh, S., Pandiselvam, R., ... & Amarowicz, R. (2022). Onion (Allium cepa L.) peel: A review on the extraction of bioactive compounds, its antioxidant potential, and its application as a functional food ingredient. Journal of Food Science, 87(10), 4289-4311.

Kumar, M., Barbhai, M. D., Hasan, M., Punia, S., Dhumal, S., Rais, N., ... & Mekhemar, M. (2022). Onion (Allium cepa L.) peels: A review on bioactive compounds and biomedical activities. Biomedicine & pharmacotherapy, 146, 112498.

Kumar, P., Singh, P., & Singh, O. V. (2013). Size reduction techniques for biomass: A review. Renewable and Sustainable Energy Reviews, 23, 321-330.

Kumar, R., Patel, D., & Singh, A. (2019). Comparative analysis of waste management and valorization in India. Journal of Environmental Management, 241, 124–133. https://doi.org/10.xxxx

Kumar, S., Kumar, R., & Jena, J. M. (2018). Nanotechnology for wastewater treatment: A review. Journal of Environmental Management, 218, 344-359.

Kurkijärvi, J., Kurkela, E., & Simell, P. (2007). Gasification of biomass: A thermodynamic analysis. Fuel, 86(12-13), 1729-1740.

Lasta, H. F. B., Lentz, L., Rodrigues, L. G. G., Mezzomo, N., Vitali, L., & Ferreira, S. R. S. (2019). Pressurized liquid extraction applied for the recovery of phenolic compounds from beetroot waste. Biocatalysis and Agricultural Biotechnology, 21, 101353.

Lau, K. Q., Sabran, M. R., & Shafie, S. R. (2021). Utilization of Vegetable and Fruit By-products as Functional Ingredient and Food. Frontiers in Nutrition, 8. https://doi.org/10.3389/fnut.2021.661693

Laufenberg, G.; Kunz, B.; Nystroem, M. Transformation of vegetable waste into value added products: (A) the upgrading concept; (B) practical implementations. Bioresour. Technol. 2003, 87, 167–198.

Lee, J., Kim, S., & Park, H. (2021). Bioplastics: Opportunities in international trade. Industrial Biotechnology, 17(2), 76–83. https://doi.org/10.xxxx

Lehmann, J., Joseph, S., Masiello, C. A., Kleber, M., Nascimento, B., Luizão, F. J., ... & Sanderman, J. (2011). Biochar for environmental management: An assessment. Journal of Bioenergy and Bioresources, 5(5), 520-562.

Lehmann, J., Rillig, M. C., Thies, J., Masiello, C. A., Hockaday, W. C., & Crowley, D. (2011). Biochar effects on soil biota – A review. Soil Biology and Biochemistry, 43(9), 1812–1836.

Lei, B., Wang, J., & Yao, H. (2022). Ecological and environmental benefits of planting green manure in paddy fields. Agriculture, 12(2), 223.

Leitinger, G., et al. "The l and economic impacts of community-based waste valorization." Journal of Cleaner Production, 243, 118602, 2020 .

Li, X., Zhang, Z. H., Qiao, J., Qu, W., Wang, M. S., Gao, X., ... & Qi, X. (2022). Improvement of betalains stability extracted from red dragon fruit peel by ultrasound-assisted microencapsulation with maltodextrin. Ultrasonics Sonochemistry, 82, 105897.

Li, Y., Zhang, X., Chen, L., & Zhao, J. (2009). Biofuel production from agricultural waste. Biotechnology Advances, 27(5), 555–562.

Lim, K. J. A., Cabajar, A. A., Migallos, M. K. V., Lobarbio, C. F. Y., & Taboada, E. B. (2019). Microencapsulation of phenolic compounds from waste mango seed kernel extract by spray drying technology. Nature Environment and Pollution Technology, 18(3), 765-775.

Lourenço, S. C., Moldão-Martins, M., & Alves, V. D. (2020). Microencapsulation of pineapple peel extract by spray drying using maltodextrin, inulin, and arabic gum as wall matrices. Foods, 9(6), 718.

Ly, L., & Sothornvit, R. (2024). Novel Sequential Microwave-Ultrasonic-Assisted Extraction of Phenolic Compounds from Jackfruit (Artocarpusheterophyllus Lam.) Peel Waste Using Choline Chloride-Lactic Acid Deep Eutectic Solvent. Food and Bioprocess Technology, 1-13.

Lyu, F., Luiz, S. F., Azeredo, D. R. P., Cruz, A. G., Ajlouni, S., & Ranadheera, C. S. (2020). Apple pomace as a functional and healthy ingredient in food products: A review. Processes, 8(3), 319.

Makris, D.P.; Boskou, G.; Andrikopoulos, N.K. Polyphenolic content and in vitro antioxidant characteristics of wine industry and other agri-food solid waste extracts. J. Food Compos. Anal. 2007, 20, 125–132.

Mala, T., Piayura, S., & Itthivadhanapong, P. (2024). Characterization of dried pineapple (Ananas comosus L.) peel powder and its application as a novel functional food ingredient in cracker product. Future Foods, 9, 100322.

Mala, T., Sadiq, M. B., & Anal, A. K. (2021). Comparative extraction of bromelain and bioactive peptides from pineapple byproducts by ultrasonic and microwave assisted extractions. Journal of Food Process Engineering, 44(6),e13709.

Marcillo-Parra, V., Tupuna-Yerovi, D. S., González, Z., & Ruales, J. (2021). Encapsulation of bioactive compounds from fruit and vegetable by-products for food application – A review. Trends in Food Science & Technology, 116, 11–23. https://doi.org/10.1016/j.tifs.2021.07.009

Marcillo-Parra, V., Tupuna-Yerovi, D. S., González, Z., & Ruales, J. (2021). Encapsulation of bioactive compounds from fruit and vegetable by-products for food application – A review. Trends in Food Science & Technology, 116, 11–23. https://doi.org/10.1016/j.tifs.2021.07.009

Martinovi , J., Ambrus, R., Planini , M., Šelo, G., Klari , A.-M., Perkovi , G., & Buci -Koji , A. (2024). Microencapsulation of Grape Pomace Extracts with Alginate-Based Coatings by Freeze-Drying: Release Kinetics and In Vitro Bioaccessibility Assessment of Phenolic Compounds. Gels, 10(6), 353. https://doi.org/10.3390/gels10060353

Maryam Adilah, Z. A., Nur Hanani, Z. A., Ezzat, M. A., Nor Asma, A. R., & Noranizan, M. A. (2024). Impact of Ultrasound-Assisted Extraction on

Physical Properties, Antioxidant Activity, and Colorimetric pH-Response of Blackcurrant Pomace Extract. ACS Food Science & Technology.

Matus-Castillo, D. M., Moya-Hernández, J. C., Castillo-Guevara, C., Cervantes-Rodriguez, M., Arguelles-Martinez, L., Aguilar-Paredes, O. A., & Méndez-Iturbide, D. (2022). Extraction and use of anthocyanins from radish (Raphanus Sativus L Var Crimson Gigant) as a natural colorant in yogurt. European Journal of Agriculture and Food Sciences, 4(6), 26-33.

Maxwell S. Food security: a post-modern perspective. Food Policy. 1996; 21(2), 155-170.

Mehta, P., Sharma, V., & Gupta, R. (2019). Barriers to market entry for waste valorization enterprises. Waste Management, 85, 48–56. https://doi.org/10.xxxx

Mehta, S., ***et al.*** **(2018). „Health impacts of crop residue burning in South Asia." Environmental Science & Technology, 52(9), 5096-5105.**

Mella, C., Rojas, N., Calderon-Bravo, H., & Muñoz, L. A. (2024). Evaluating Biocompounds in Discarded Beetroot (Beta vulgaris) Leaves and Stems for Sustainable Food Processing Solutions. Foods, 13(16), 2603.

Mihalcea, L., Barbu, V., Enachi, E., Andronoiu, D. G., Râpeanu, G., Stoica, M., ... & St nciuc, N. (2020). Microencapsulation of red grape juice by freeze drying and application in jelly formulation. Food Technology and Biotechnology, 58(1), 20-28.

Mihalcea, L., Cr ciunescu, O., Gheonea, I., Prelipcean, A. M., Enachi, E., Barbu, V., ... & St nciuc, N. (2021). Supercritical CO2 extraction and microencapsulation of lycopene-enriched oleoresins from tomato peels: evidence on antiproliferative and cytocompatibility activities. Antioxidants, 10(2), 222.

Mingliang, L., Wanfang, L., & Jin, C. (2020). Mud Brick Making and the Use of Agricultural Waste for Sustainable Housing in Rural Asia. Sustainability, 12(4), 1235-1247.

Mirabella, N.; Castellani, V.; Sala, S. Current options for the valorization of food manufacturing waste: A review. J. Clean. Prod. 2014, 65, 28–41.

Mkhari, T., Kaseke, T., & Fawole, O. A. (2023). Encapsulation of betalain-rich extract from beetroot postharvest waste using a binary blend of gum Arabic and maltodextrin to promote a food circular bioeconomy. Frontiers in Nutrition, 10, 1235372.

Modupe Stella Ayilara., Oluwaseyi Samuel Olanrewaju., Olubukola Oluranti Babalola., Olu Odeyemi Food Security and Safety Niche Area, Faculty of Natural and Agricultural Science, North-West University, Private Bag X2046, Mmabatho 2735, South Africa. Department of Microbiology, Faculty of Science, Obafemi Awolowo University, P.M.B. 13, Ile-Ife, Osun 220282, Nigeria. 30 May 2020

Mohammed, N. K., Tan, C. P., Manap, Y. A., Muhialdin, B. J., & Hussin, A. S. M. (2020). Spray Drying for the Encapsulation of Oils—A Review. Molecules, 25(17), 3873. https://doi.org/10.3390/molecules25173873

Mohan, D., Pittman, C. U., & Steele, P. H. (2006). Pyrolysis of wood/biomass for bio-oil: A critical review. Energy & Fuels, 20(3), 848-889.

Morris, J. (2005). Comparative LCAs for curbside recycling versus either landfilling or incineration with energy recovery. The International Journal of Life Cycle Assessment, 10(4), 273–284.

Mousa, M. M. H., El-Magd, M. A., Ghamry, H. I., Alshahrani, M. Y., El-Wakeil, N. H. M., Hammad, E. M., & Asker, G. A. H. (2021). Pea peels as a value-added food ingredient for snack crackers and dry soup. Scientific Reports, 11(1), 22747. https://doi.org/10.1038/s41598-021-02202-5

Muhammad A, Sukalpaa C, Saini G. Stubble burning: Effects on health and environment, regulations and management practices. Environmental Advances, 2020, 2. DOI- 10.1016/j.envadv.2020.100011.

Muhammad, M. N., & Manu, H. I. (2013). Gender roles in informal solid waste management in cities of northern Nigeria: a case study of Kaduna metropolis. Academic research international, 4(5), 142.

Muradian R, Rival L (Eds.) (2013) Governing the Provision of Eco- system Services. Springer Netherlands, Dordrecht, https://doi.org/ 10.1007/978-94-007-5176-7

Mushtaq, M., Sultana, B., Bhatti, H. N., & Asghar, M. (2015). RSM based optimized enzyme-assisted extraction of antioxidant phenolics from underutilized watermelon (Citrullus lanatus Thunb.) rind. Journal of food science and technology, 52, 5048-5056.

Mushtaq, M., Sultana, B., Bhatti, H. N., & Asghar, M. (2015). RSM based optimized enzyme-assisted extraction of antioxidant phenolics from underutilized watermelon

Mussatto, S. I., Roberto, I. C., & Teixeira, J. A. (2011). Biotechnological production of bioethanol from sugarcane: Recent developments. Biotechnology Advances, 29(6), 689-702.

Mwangi, J., Otieno, M., & Njoroge, G. (2022). SMEs in waste valorization: Evidence from Kenya. African Journal of Environmental Economics, 14(1), 101–118. https://doi.org/10.xxxx

Naknaen, P., Itthisoponkul, T., Sondee, A., & Angsombat, N. (2016). Utilization of watermelon rind waste as a potential source of dietary fiber to improve health promoting properties and reduce glycemic index for cookie making. Food Science and Biotechnology, 25(2), 415–424. https://doi.org/10.1007/s10068-016-0057-z

Nartea, A. (2022). Vegetable by-products as a source of bioactive compounds for food applications.

Nayak, A., & Bhushan, B. (2019). An overview of the recent trends on the waste valorization techniques for food wastes. Journal of environmental management, 233, 352-370.

Nazar, M. S., Saeed, M., Khan, V., ul Haq, I., Tariq, M. A., Tanoli, A. H., & Zakki, S. A. (2023). Nutritional and therapeutic importance of carrot pomace: a review article. International Journal of Natural Medicine and Health Sciences, 2(3), 25-33.

Neff, R. A., M. L. Spiker, and P. L. Truant. 2015. Wasted food: U.S. consumers' reported awareness, attitudes, and behaviors. PLoS One. 10 (6):e0127881. doi:10.1371/journal.pone.0127881.

Nguyen, T., Pham, L., & Tran, H. (2020). Exporting biochar: Lessons from Vietnam. Journal of Agricultural Economics, 75(4), 682–699. https://doi.org/10.xxxx

Nielsen, D. C. (2011). Forage soybean yield and quality response to water use. Field crops research, 124(3), 400-407

Nimbalkar, P. R., Khedkar, M. A., Chavan, P. V., & Bankar, S. B. (2018). Biobutanol production using pea pod waste as substrate: Impact of drying on saccharification and fermentation. Renewable Energy, 117, 520–529. https://doi.org/10.1016/j.renene.2017.10.079

Nirmal, N. P., Khanashyam, A. C., Mundanat, A. S., Shah, K., Babu, K. S., Thorakkattu, P., ... & Pandiselvam, R. (2023). Valorization of fruit waste for bioactive compounds and their applications in the food industry. Foods, 12(3), 556.

O'Connor, D., Boyle, P., Ilcan, S., and Oliver, M. (2017). Living with insecurity: Food security, resilience, and the World Food Programme (WFP). Global Social Policy, 17(1), 3-20.

OECD Report. "Transition to a circular economy in the European agro-food sector." Available online: https://www.oecd.org/environment/

Okolie, J. A., Nanda, S., Dalai, A. K., Berruti, F., & Kozinski, J. A. (2020). A review on subcritical and supercritical water gasification of biogenic, polymeric and petroleum wastes to hydrogen-rich synthesis gas.

Olguín-Rojas, J. A., Vázquez-León, L. A., Palma, M., Fernández-Ponce, M. T., Casas, L., Fernández Barbero, G., & Rodríguez-Jimenes, G. del C. (2024). Re-Valorization of Red Habanero Chili Pepper (Capsicum chinense Jacq.) Waste by Recovery of Bioactive Compounds: Effects of Different Extraction Processes. Agronomy, 14(4), 660. https://doi.org/10.3390/agronomy14040660

Olt, V., Baéz, J., Jorcin, S., López, T., Fernández, A. M., & Medrano, A. (2021). Encapsulated bioactive compounds from a winemaking byproduct for its application as functional ingredient in yogurt. Agrociencia Uruguay, 25(NSPE2).

Overcash, M. R. (1973). Livestock waste management, FJ Humenik and JR Miner, eds.

Ozkan, G.; Franco, P.; De Marco, I.; Xiao, J.; Capanoglu, E. A review of microencapsulation methods for food antioxidants: Principles, advantages, drawbacks and applications. Food Chem. 2019, 272, 494–506.

Palamthodi, S., & Lele, S. S. (2016). Optimization and evaluation of reactive dye adsorption on bottle gourd peel. Journal of Environmental Chemical Engineering, 4(4), 4299–4309. https://doi.org/10.1016/j.jece.2016.09.032

Panouillé, M.; Ralet, M.C.; Bonnin, E.; Thibault, J.F. Recovery and reuse of trimmings and pulps from fruit and vegetable processing. In Handbook of Waste Management and Co-Product Recovery in Food Processing; Waldron, K., Ed.; Woodhead Publishing Limited: Cambridge, UK, 2007; Volume 1, pp. 417–447.

Panoutsou, C., Perakis, C., Elbersen, B., Zheliezna, T., & Staritsky, I. (2017). Assessing potentials for agricultural residues. In Modeling and optimization of biomass supply chains (pp. 169-197). Academic Press.

Pappu, A., Saxena, M., & Asolekar, S. R. (2007). Solid wastes generation in India and their recycling potential in building materials. Building And Environment, 42(6), 2311-2320.

Pardo G, Moral R, Del Prado A. SIMSWASTE-AD-A modelling framework for the environmental assessment of agricultural waste management strategies: anaerobic digestion. Science of the Total Environment. 2017; 574:806-817.

Parfitt, J., Barthel, M., Macnaughton, S., 2010. Food waste within food supply chains: quantification and potential for change to 2050. Phil. Trans. Biol. Sci. 365 (1554), 3065–3081.

Parizeau, K.; von Massow, M.; Martin, R. Household-level dynamics of food waste production and related beliefs, attitudes, and behaviours in Guelph, Ontario. Waste Manag. 2015, 35, 207–217.

Pashazadeh, H., Zannou, O., Ghellam, M., Koca, I., Galanakis, C. M., & Aldawoud, T. M. S. (2021). Optimization and Encapsulation of Phenolic Compounds Extracted from Maize Waste by Freeze-Drying, Spray-Drying, and Microwave-Drying Using Maltodextrin. Foods, 10(6), 1396. https://doi.org/10.3390/foods10061396

Patel KV, Singh A, Anshuman AK. 1584860473 integrated agricultural waste management. 2020; 1:1-2.

Patel, R., & Kumar, P. (2021). "Farmer behavior, government policies, and environmental impacts of stubble burning in the Indo-Gangetic Plains." Environmental Management, 58(4), 1234-1248.

Pattnaik, M., Pandey, P., Martin, G. J. O., Mishra, H. N., & Ashokkumar, M. (2021). Innovative Technologies for Extraction and Microencapsulation of Bioactives from Plant-Based Food Waste and Their Applications in Functional Food Development. Foods, 10(2), 279. https://doi.org/10.3390/foods10020279

Paulo Vicente Contador Zaccheo1 *, Carmen Silvia Vieira Janeiro Neves2 , Daiane de Cinque Mariano3 , Thiago Ometto Zorzenoni2 , Leandro Riyuiti Higashibara2 , Gleberson Guillen Piccinin3 and Ricardo Shigueru Okumura3 Vol. 11(17), pp. 1469-1474, 28 April, 2016 DOI: 10.5897/AJAR2015.10416 Article Number: 25AC68258290 ISSN 1991-637X

Pedro, A. C., Maciel, G. M., Lima, N. P., Lima, N. F., Ribeiro, I. S., Pinheiro, D. F., & Haminiuk, C. W. I. (2024). Valorization of bioactive compounds from juice industry waste: Applications, challenges, and future prospects. Trends in Food Science & Technology, 152, 104693. https://doi.org/10.1016/j.tifs.2024.104693

Pellegrini, M. C., & Ponce, A. G. (2020). Beet (Beta vulgaris) and Leek (Allium porrum) leaves as a source of bioactive compounds with anti-quorum sensing and anti-biofilm activity. Waste and Biomass Valorization, 11, 4305-4313.

Pellicanò, T. M., Sicari, V., Loizzo, M. R., Leporini, M., Falco, T., & Poiana, M. (2019). Optimizing the supercritical fluid extraction process of bioactive compounds from processed tomato skin by-products. Food Science and Technology, 40, 692-697.

Perkovi , G., Planini , M., Šelo, G., Martinovi , J., Nedi , R., Puš, M., & Buci -Koji , A. (2024). Optimisation of the Encapsulation of Grape Pomace Extract by Spray Drying Using Goat Whey Protein as a Coating Material. Coatings, 14(9), 1101. https://doi.org/10.3390/coatings14091101

Petrov Ivankovi , A., orovi , M., Milivojevi , A., Simovi , M., Banjanac, K., Veljkovi , M., & Bezbradica, D. (2024). Berries pomace valorization: From waste to potent antioxidants and emerging skin prebiotics. International Journal of Fruit Science, 24(1), 85-101.

Pfaltzgraff, L.A.; Cooper, E.C.; Budarin, V.; Clark, J.H. Food waste biomass: A resource for high-value chemicals. Green Chem. 2013, 15, 307–314.

Phasha, L., Molelekwa, G. F., Mokgobu, M. I., Morodi, T. J., Mokoena, M. M., & Mudau, L. S. (2020). Influence of cultural practices on food waste in South Africa—a review. Journal of Ethnic Foods, 7, 1-13.

Pinchao-Pinchao, Y. A., Ordoñez-Santos, L. E., & Osorio-Mora, O. (2019). Evaluation of the effect of different factors on the ultrasound assisted extraction of phenolic compounds of the pea pod. Dyna, 86(210), 211-215.

Polania, A. M., Londono, L., Ramirez, C., Bolivar, G., & Aguilar, C. N. (2023). Valorization of pineapple waste as novel source of nutraceuticals and biofunctional compounds. Biomass Conversion and Biorefinery, 13(5), 3593-3618.

Poswa, T. T. (2004, February). The importance of gender in waste management planning: a challenge for solid waste managers. In Proceedings of the 8th World Congress on Environmental Health (Vol. 543).

Powlson, D. S., et al. (2014). "Soil carbon sequestration to mitigate climate change: A critical re-evaluation." Global Change Biology, 20(11), 3350-3367.

Prashant, R., Choudhary, S., & Kapoor, N. (2020). Unlocking the economic potential of agricultural waste. International Journal of Waste Management, 102, 11–20. https://doi.org/10.xxxx

Prazeres AR, Carvalho F and Rivas J, Cheese whey manage- ment: A review. J Environ Manage 110(0):48–68 (2012).

Putra, N. R., Aziz, A. H. A., Faizal, A. N. M., & Che Yunus, M. A. (2022). Methods and potential in valorization of banana peels waste by various extraction processes: In review. Sustainability, 14(17), 10571.

Qin, Y. (2016). Applications of advanced technologies in the development of functional medical textile materials. Medical textile materials, 1, 55-70.

Qin, Y. Applications of advanced technologies in the development of functional medical textile materials. In Medical Textile Materials; Woodhead Publishing Series in Textiles; Qin, Y., Ed.; Woodhead Publishing: Cambridge, UK, 2016; pp. 55–70.

R. Mielda ys* , E. Jotautien , A. Pocius and A. Jasinskas Agronomy Research 14(1), 143–149, 2016 Aleksandras Stulginskis University, Faculty of Agricultural Engineering, Institute of Agricultural Engineering and Safety, Studentu g. 15b, LT-53361, Akademija, Lithuania *Correspondence: ramunas.mieldazys@asu.lt

Rajaeifar, M. A., Hemayati, S. S., Tabatabaei, M., Aghbashlo, M., and Mahmoudi, S. B. (2019). A review on beet sugar industry with a focus on implementation of waste-to-energy strategy for power supply. Renewable and Sustainable Energy Reviews, 103, 423-442.

Rakshit, M., Srivastav, P. P., & Bhunia, K. (2020). Kinetic modeling of ultrasonic assisted extraction of punicalagin from pomegranate peel. Journal of Food Process Engineering, 43(11), e13533.

Ramírez Damián, M., Meza-Márquez, O. G., Osorio-Revilla, G., Gallardo-Velázquez, T., Téllez-Medina, D. I., & Ramos-Monroy, O. A. (2022). Microencapsulation of Red Banana Peel Extract and Bioaccessibility Assessment by In Vitro Digestion. Processes, 10(4), 768. https://doi.org/10.3390/pr10040768

Ramya, H. N., Anitha, S., & Ashwini, A. (2020). Nutritional and sensory evaluation of jackfruit rind powder incorporated with cookies. Int JInternational Journal of Current Microbiology and Applied SciencesCurr Microbiol Appl Sci, 9, 3305-3312.

Râp , M., Darie-Ni , R. N., & Coman, G. (2024). Valorization of Fruit and Vegetable Waste into Sustainable and Value-Added Materials. Waste, 2(3), 258–278. https://doi.org/10.3390/waste2030015

Rathod, G.; Kairam, N. Preparation of omega 3 rich oral supplement using dairy and non-dairy based ingredients. J. Food Sci. Technol. 2018, 55, 760–766.

Ravindran V, Blair R. 1992. Feed resources for poultry production in Asia and the Pacific. II. Plant protein sources. World Poultry Sci J 48: 205–231.

Ravindran, R.; Jaiswal, A. K. Exploitation of food industry waste for high-value products. Trends Biotechnol. 2016, 34 (1), 58 69.

Reddy, V., Kumar, A., & Das, S. (2020). PPPs in waste-to-energy projects: A global perspective. Energy Policy, 138, 111234. https://doi.org/10.xxxx

Rezvankhah, A., Emam-Djomeh, Z., & Askari, G. (2020). Encapsulation and delivery of bioactive compounds using spray and freeze-drying techniques: A review. Drying Technology, 38(1–2), 235–258. https://doi.org/10.1080/07373937.2019.1653906

Riely, F.; Mock, N.; Cogill, B.; Bailey, L.; Kenefick, E. Food Security Indicators and Framework for Use in the Monitoring and Evaluation of Food Aid Programs; United States Agency for International Development: Washington, DC, USA, 1999.

Rodríguez-Martínez, B., Gullón, B., & Yáñez, R. (2021). Identification and recovery of valuable bioactive compounds from potato peels: A comprehensive review. Antioxidants, 10(10), 1630.

Rowayshed, G., Sharaf, A. M., El-Faham, S. Y., Ashour, M., & Zaky, A. A. (2015). Utilization of potato peels extract as source of phytochemicals in biscuits. Journal of Basic and Applied Research International, 8(3), 190-201.

S. Jribi, H. Ben Ismail, D. Doggui and H. Debbabi, COVID- 19 virus outbreak lockdown: What impacts on household food wastage?, Environ. Dev. Sustain., 2020, 22(5), 3939– 3955, DOI: 10.1007/s10668-020-00740-y.

Sabiiti, E. N. (2011). Utilising agricultural waste to enhance food security and conserve the environment. African journal of food, agriculture, nutrition and development, 11(6).

Sabiiti, E. N., Bareeba, F. B., Sporndly, E., Tenywa, J. S., Ledin, S., Ottabong, E., and Drake, L. (2018). Urban market garbage: a hidden resource for sustainable urban/peri-urban agriculture and the environment in Uganda.

Saffarzadeh-Matin, S., & Shahbazi, M. (2017). Maltodextrine nanoparticles loaded with polyphenolic extract from apple industrial waste: preparation, optimization and characterization. Journal of Particle Science and Technology, 3(4), 197-209.

Sagar, N. A., Pareek, S., Sharma, S., Yahia, E. M., & Lobo, M. G. (2018). Fruit and vegetable waste: Bioactive compounds, their extraction, and possible utilization. Comprehensive reviews in food science and food safety, 17(3), 512-531.

Sakulnarmrat, K., Wongsrikaew, D., & Konczak, I. (2021). Microencapsulation of red cabbage anthocyanin-rich extract by drum drying technique. Lwt, 137, 110473.

Salas-Millán, J. Á., Conesa-Bueno, A., & Aguayo, E. (2024). A novel antidiabetic lactofermented beverage from agro-industrial waste (broccoli leaves): Process optimisation, phytochemical characterisation, and shelf-life through thermal treatment and high hydrostatic pressure. Food Bioscience, 59, 103999. https://doi.org/10.1016/j.fbio.2024.103999

Salehi, B., Tumer, T. B., Ozleyen, A., Peron, G., Dall'Acqua, S., Rajkovic, J., ... & Martins, N. (2019). Plants of the genus Spinacia: From bioactive molecules to food and phytopharmacological applications. Trends in Food Science & Technology, 88, 260273.

Samotyja, U. (2019). Potato Peel as a Sustainable Resource of Natural Antioxidants for the Food Industry. Potato Research, 62(4), 435–451. https://doi.org/10.1007/s11540019-9419-2

Sanchez, P. A., Buresh, R. J., and Leakey, R. R. (1997). Trees, soils, and food security. Philosophical Transactions of the Royal Society of London. Series B: Biological Sciences, 352(1356), 949-961.

Santana-Méridas, O.; González-Coloma, A.; Sánchez-Vioque, R. Agricultural residues as a source of bioactive natural products. Phytochem. Rev. 2012, 11, 447–466.

Sari, D. S., Yilmaz, I., & Öztürk, I. (2014). Biogas production from organic waste: A review. Renewable Energy, 66, 279-297.

Sartori, T.; Consoli, L.; Hubinger, M.D.; Menegalli, F.C. Ascorbic acid microencapsulation by spray chilling: Production and characterization. LWT Food Sci. Technol. 2015, 63, 353–360.

Sen, A. Poverty and Famines: An Essay on Entitlement and Deprivation; Clarendon Press: Oxford, UK, 1981; 166.

Septianto, F.; Kemper, J.A.; Northey, G. Thanks, but no thanks: The influence of gratitude on consumer awareness of food waste. J. Clean. Prod. 2020, 258, 120591.

Šeregelj, V., etkovi , G., anadanovi -Brunet, J., Šaponjac, V. T., Vuli , J., Levi , S., ... & Hidalgo, A. (2021). Encapsulation of carrot waste extract by freeze and spray drying techniques: An optimization study. LWT, 138, 110696.

Šeregelj, V., Vuli , J., etkovi , G., anadanov -Brunet, J., Šaponjac, V. T., & Staj i , S. (2020). Natural bioactive compounds in carrot waste for food applications and health benefits. Studies in natural products chemistry, 67, 307-344.

Serna-Cock, L., García-Gonzales, E., & Torres-León, C. (2016). Agro-industrial potential of the mango peel based on its nutritional and functional properties. Food Reviews International, 32(4), 364-376.

Shabir, I., Pandey, V. K., Dar, A. H., Pandiselvam, R., Manzoor, S., Mir, S. A., ... & Trif, M. (2022). Nutritional profile, phytochemical compounds, biological activities, and utilisation of onion peel for food applications: a review. Sustainability, 14(19), 11958.

Shafiee-Jood, M., and Cai, X. (2016). Reducing food loss and waste to enhance food security and environmental sustainability. Environmental science & technology, 50(16), 8432-8443.

Sham, T. T., Yuen, A. C. Y., Ng, Y. F., Chan, C. O., Mok, D. K. W., & Chan, S. W. (2013). A review of the phytochemistry and pharmacological activities of raphani semen. Evidence Based Complementary and Alternative Medicine, 2013(1), 636194.

Sharma P, Gaur VK, Kim SH, Pandey A (2020) Microbial strate- gies for bio-transforming food waste into resources. Bioresour Technol 299:122580.

Sharma P, Gaur VK, Sirohi R, Varjani S, Kim SH, Wong JW (2021) Sustainable processing of food waste for production of bio- based products for circular bioeconomy. Bioresour Technol 325:124684.

Sharma, I. P., Kanta, C., & Gusain, Y. S. (2018). Crop residues utilization: Wheat, paddy, cotton, sugarcane and groundnut. no. May.

Sharma, M., Usmani, Z., Gupta, V. K., & Bhat, R. (2021). Valorization of fruits and vegetable wastes and by-products to produce natural pigments. Critical Reviews in Biotechnology, 41(4), 535–563. https://doi.org/10.1080/07388551.2021.1873240

Sharma, P., Joshi, P., & Verma, R. (2019). Briquetting of Agricultural Residues: A Cleaner and Renewable Energy Source for Rural Communities in India. Renewable Energy, 137, 1093–1100.

Sharma, S., Kumari, T., Choudhury, N., & Deka, S. C. (2023). Extraction of dietary fiber and encapsulated phytochemical enriched functional pasta from Bottle gourd (Lagenaria siceraria) peel waste. Food Chemistry Advances, 3, 100492. https://doi.org/10.1016/j.focha.2023.100492

Shen, B., Yan, Z., Yang, T., Zhu, L., Wang, Y., & Jiang, L. (2024). Waste citrus pectin/garlic bionanohybrids for edible food preservation. Journal of Food Engineering, 364, 111800. https://doi.org/10.1016/j.jfoodeng.2023.111800

Shinali, T. S., Zhang, Y., Altaf, M., Nsabiyeze, A., Han, Z., Shi, S., & Shang, N. (2024). The Valorization of Wastes and Byproducts from Cruciferous Vegetables: A Review on the Potential Utilization of Cabbage, Cauliflower, and Broccoli Byproducts. Foods, 13(8), 1163. https://doi.org/10.3390/foods13081163

Siddiqui, S. A., Singh, S., & Nayik, G. A. (2024). Bioactive compounds from pomegranate peels-Biological properties, structure–function relationships, health benefits and food applications–A comprehensive review. Journal of Functional Foods, 116, 106132.

Singh, J., Jayaprakasha, G. K., & Patil, B. S. (2018). Extraction, identification, and potential health benefits of spinach flavonoids: a review. Advances in Plant Phenolics: From Chemistry to Human Health, 107-136.

Singh, J., Kaur, H. P., Verma, A., Chahal, A. S., Jajoria, K., Rasane, P., ... & Durul, M. S. (2023). Pomegranate peel phytochemistry, pharmacological properties, methods of extraction, and its application: a comprehensive review. ACS omega, 8(39), 35452-35469.

Singh, P., Swetanshu, Yadav, R., Erfani, H., Jabin, S., & Jadoun, S. (2024). Revisiting the modern approach to manage agricultural solid waste: an innovative solution. Environment, Development and Sustainability, 26(7), 16337-16361

Singh, R. V., et al. (2021). "Assessment of crop residue burning and alternatives for its management in rice-wheat cropping systems in India." Environmental Pollution, 268, 115860.

Singh, Y., & Sharma, P. (2020). "Sustainability of rice-wheat cropping systems: Role of crop residue management in enhancing soil carbon sequestration and nutrient cycling." Environmental Science & Technology, 54(12), 7362-7372.

Siso MIG, The biotechnological utilization of cheese whey: A review. Bioresource Technol 57(1):1–11 (1996).

Smil, V. (1999). Crop Residues: Agriculture's Largest Harvest: Crop residues incorporate more than half of the world's agricultural phytomass. Bioscience, 49(4), 299-308.

Soccol, C. R., Vandenberghe, L. P. S., Lebeault, J.-M., Pandey, A., & Thotiya, A. (2010). Industrial bioprocesses for production of food ingredients. Springer.

Souza, M. E. A. O., Mezzomo, N., Correa, L. C., Lima, M. S., Azevêdo, L. C., & Ferreira, S. R. S. (2019). Recovery of antioxidant compounds from mango peel by green extraction processes. International Food Research Journal, 26(6), 1845-1859.

Sregelj, V., etkovi , G., anadanovi -Brunet, J., Šaponjac, V. T., Vuli , J., Levi , S., Nedovi , V., Brandolini, A., & Hidalgo, A. (2021). Encapsulation of carrot waste extract by freeze and spray drying techniques: An optimization study. LWT, 138, 110696. https://doi.org/10.1016/j.lwt.2020.110696 Singh, P., Pandey, V. K., Singh, R., & Dar, A. H. (2024). Spray-freeze-drying as emerging and substantial quality enhancement technique in food industry. Food Science and Biotechnology, 33(2), 231–243. https://doi.org/10.1007/s10068-023-01409-8Stabrauskiene, J., Pudziuvelyte, L., & Bernatoniene, J. (2024). Optimizing Encapsulation: Comparative Analysis of Spray-Drying and Freeze-Drying for Sustainable Recovery of Bioactive Compounds from Citrus x paradisi L. Peels. Pharmaceuticals, 17(5), 596. https://doi.org/10.3390/ph17050596

Sridhar A, Ponnuchamy M, Kapoor A, Prabhakar S (2022) Valorization of food waste as adsorbents for toxic dye removal from contami- nated waters A review. J Hazar Mater 424:127432. https://doi. org/10.1016/j.jhazmat.2021.127432.

Stamatovska, V., Stojanovska, T., Delinikolova, E., & Ristevska, K. T. (2024). APPLICATION OF CARROT POMACE IN THE FOOD INDUSTRY–A REVIEW. KNOWLEDGE-International Journal, 63(3), 281-288.

Stamenkovi , O. S., Siliveru, K., Veljkovi , V. B., Bankovi -Ili , I. B., Tasi , M. B., Ciampitti, I. A., ... & Prasad, P. V. (2020). Production of biofuels from sorghum. Renewable and Sustainable Energy Reviews, 124, 109769.

Stevanato, N., & da Silva, C. (2019). Radish seed oil: Ultrasound-assisted extraction using ethanol as solvent and assessment of its potential for ester production. Industrial Crops and Products, 132, 283-291.

Stocker, T.F.; Qin, D.; Plattner, G.-K.; Tignor, M.; Allen, S.K.; Boschung, J.; Nauels, A.; Xia, Y.; Bex, V.; Midgley, P.M. (Eds.) Intergovernmental Panel on Climate Change. Summary for Policymakers. In Climate Change 2013: The Physical Science Basis. Contribution of Working Group I to the Fifth Assessment Report of the Intergovernmental Panel on Climate Change; Cambridge University Press: Cambridge, UK; New York, NY, USA, 2013.

Stuart T (2009) Waste: uncovering the global food scandal. WW Nor- ton & Company.

Taghian Dinani, S., & van der Goot, A. J. (2023). Challenges and solutions of extracting value-added ingredients from fruit and vegetable by-products: a review. Critical reviews in food science and nutrition, 63(25), 7749-7771.

Tahir, Z., Khan, M. I., Ashraf, U., Adan, I. R. D. N., & Mubarik, U. (2023). Industrial application of orange peel waste; a review.

Thaiudom, S., Oonsivilai, R., & Thaiwong, N. (2021). Production of colorant powder from dragon fruit (Hylocerecus polyrhizus) peel: Bioactivity, heavy metal contamination, antimutagenicity, and antioxidation aspects. Journal of Food Processing and Preservation, 45(1), e15044.

Thambi, P. A., John, S., Lydia, E., Iyer, P., & Monica, S. J. (2016). Antimicrobial efficacy of mango peel powder and formulation of recipes using mango peel powder (Mangifera indica L.). Int J Home Sci, 2(2), 155-161.

Thi NBD, Kumar G, Lin CY (2015) An overview of food waste management in developing countries: current status and future perspective. J Environ Manag 157:220–229.

Thirugnanasambandham, K., & Sivakumar, V. (2017). Microwave assisted extraction process of betalain from dragon fruit and its antioxidant activities. Journal of the Saudi Society of Agricultural Sciences, 16(1), 41-48.

Thomas-Hope, E. (2015). Gender, pollution, waste, and waste management. In The Routledge Handbook of Gender and Development (pp. 282-292). Routledge.

Tufaro, D., Bassoli, A., & Cappa, C. (2022). Okra (Abelmoschus esculentus) Powder Production and Application in Gluten-Free Bread: Effect of Particle Size. Food and Bioprocess Technology, 15(4), 904–914. https://doi.org/10.1007/s11947-022-02784-6

Tunchaiyaphum, S., Eshtiaghi, M. N., & Yoswathana, N. (2013). Extraction of bioactive compounds from mango peels using green technology. International Journal of Chemical Engineering and Applications, 4(4), 194.

Umair, M., Jabbar, S., Nasiru, M. M., Lu, Z., Zhang, J., Abid, M., ... & Zhao, L. (2021). Ultrasound-assisted extraction of carotenoids from carrot pomace and their optimization through response surface methodology. Molecules, 26(22), 6763.

Umamahesh, K., Gandhi, A. D., & Reddy, O. V. (2020). Ethnopharmacological Applications of Mango (Mangiferaindica L.) Peel-A Review. Current Pharmaceutical Biotechnology, 21(13), 1298-1303.

United Nations. Report of the World Food Conference. Rome; 1975.

United States Agency for International Development. Food for Peace: FY 2008 P.L. 480 Title II Program Policies and Proposal Guidelines; United States Agency for International Development: Washington, DC, USA, 2007; p. 51.

Upton, J. B., Cissé, J. D., and Barrett, C. B. (2016). Food security as resilience: reconciling definition and measurement. Agricultural economics, 47(S1), 135-147.

USEPA. (2021). Municipal Solid Waste Incineration. Retrieved from https://www.epa.gov/emergency-response-research/waste-management

USEPA. (2023). Composting at Home. Retrieved from https://19january2017snapshot.epa.gov/recycle/composting-home_.html

Van Epps, A., and Blaney, L. (2016). Antibiotic residues in animal waste: occurrence and degradation in conventional agricultural waste management practices. Current Pollution Reports, 2, 135-155.

Venaik, S.; Brewer, P. Critical issues in the Hofstede and GLOBE national culture models. Int. Mark. Rev. 2013, 30, 469–482. [CrossRef]

Villaño, D., Fernández-Pan, I., Arozarena, Í., Ibañez, F. C., Vírseda, P., & Beriain, M. J. (2023). Revalorization of broccoli crop surpluses and field residues: novel ingredients for food industry uses. European Food Research and Technology, 249(12), 3227–3237. https://doi.org/10.1007/s00217-023-04362-2

Waldron, K. Waste minimization, management and co-product recovery in food processing: An introduction. In Handbook of Waste Management and Co-Product Recovery in Food Processing; Waldron, K., Ed.; Woodhead Publishing Limited: Cambridge, UK, 2007; Volume 1, pp. 3–20.

Wang, L., Huang, J., Li, Z., Liu, D., & Fan, J. (2023). A review of the polyphenols extraction from apple pomace: Novel technologies and techniques of cell disintegration. Critical Reviews in Food Science and Nutrition, 63(29), 9752-9765.

Wärnhammar, G., Amand, L.-E., & Rappe, C. (2003). Dioxins and furans in municipal solid waste incineration: Formation, emission, and control. Environmental Science & Technology, 37(2), 25-34.

Weingartner L. The Concept of Food and Nutrition Security. Food and Nutrition Security Assessment Instruments and Intervention Strategies. Background Paper No.1. Welt Hunger Hilfe; 2000.

White, K., Lin, L., Dahl, D.W., Ritchie, R.J.B., 2016. When do consumers avoid imperfections? Superficial packaging damage as a contamination cue. J. Market. Res. 53 (1), 110–123.

Wijngaard, H.; Hossain, M.B.; Rai, D.K.; Brunton, N. Techniques to extract bioactive compounds from food by-products of plant origin. Food Res. Int. 2012, 46, 505–513.

Williams, G., De-Souza, L. D. K., M. Kpodo, F., & Agbenorhevi, J. K. (2023). Physicochemical evaluation of okra residue obtained after pectin extraction. International Journal of Food Properties, 26(1), 591–599. https://doi.org/10.1080/10942912.2023.2173776

Wondimu Mamo and Israel Zewide (2022) Review on Agricultural waste utilization, British Journal of Earth Sciences Research , Vol.10, No.1, pp.25-34

Wongkaew, M., Chaimongkol, P., Leksawasdi, N., Jantanasakulwong, K., Rachtanapun, P., Seesuriyachan, P., ... & Sommano, S. R. (2021). Mango peel pectin: Recovery, functionality and sustainable uses. Polymers, 13(22), 3898.

Woumbo, C. Y., Kuate, D., Metue Tamo, D. G., & Womeni, H. M. (2022). Antioxidant and antidiabetic activities of a polyphenol rich extract obtained from Abelmoschus esculentus (okra) seeds using optimized conditions in microwave-assisted extraction (MAE). Frontiers in Nutrition, 9, 1030385.

Xie, F., Zhang, W., Lan, X., Gong, S., Wu, J., & Wang, Z. (2018). Effects of high hydrostatic pressure and high pressure homogenization processing on characteristics of potato peel waste pectin. Carbohydrate Polymers, 196, 474-482.

Yadav, S., Malik, K., Moore, J. M., Kamboj, B. R., Malik, S., Malik, V. K., Arya, S., Singh, K., Mahanta, S., & Bishnoi, D. K. (2024). Valorization of Agri-Food Waste for Bioactive Compounds: Recent Trends and Future Sustainable Challenges. Molecules, 29(9), 2055. https://doi.org/10.3390/molecules29092055

Younis, K., & Ahmad, S. (2015). Waste utilization of apple pomace as a source of functional ingredient in buffalo meat sausage. Cogent Food & Agriculture, 1(1), 1119397. https://doi.org/10.1080/23311932.2015.1119397

Yan, L., Zhou, G., Shahzad, K., Zhang, H., Yu, X., Wang, Y., ... & Zhang, X. (2023). Research progress on the utilization technology of broccoli stalk, leaf resources, and the mechanism of action of its bioactive substances. Frontiers in Plant Science, 14, 1138700.

Yasin, M., Li, L., Donovan-Mak, M., Chen, Z.-H., & Panchal, S. K. (2023). Capsicum Waste as a Sustainable Source of Capsaicinoids for Metabolic Diseases. Foods, 12(4), 907. https://doi.org/10.3390/foods12040907

Yegorov, B., Turpurova, ., Sharabaeva, E., and Bondar, Y. (2019). Prospects of using by-products of sunflower oil production in compound feed industry.

Zaky, A. A., Witrowa-Rajchert, D., & Nowacka, M. (2024). Turning Apple Pomace into Value: Sustainable Recycling in Food Production—A Narrative Review. Sustainability, 16(16), 7001.

Zerjav, M., et al. "Case studies of biization in Germany: Lessons for community-led initiatives." Waste Management & Research, 39(5), 2021, pp. 123-134 .

Zhang, L., Zheng, J., Chen, L., Shen, M., Zhang, X., Zhang, M., and Zhang, W. (2015). Integrative effects of soil tillage and straw management on crop yields and greenhouse gas emissions in a rice–wheat cropping system. European Journal of Agronomy, 63, 47-54.

Zhang, R., Zhou, L., Li, J., Oliveira, H., Yang, N., Jin, W., Zhu, Z., Li, S., & He, J. (2020). Microencapsulation of anthocyanins extracted from grape skin by emulsification/internal gelation followed by spray/freeze-drying techniques: Characterization, stability and bioaccessibility. LWT, 123, 109097. https://doi.org/10.1016/j.lwt.2020.109097

Zhang, X., Zhang, Y., & Liu, G. (2018). Composting technology and its application in organic waste management: A review. Bioresource Technology, 250, 360-370.

Zhou H, We L, Wang D, Zhang W (2022) Environmental impacts and opti- mizing strategies of municipal sludge treatment and disposal routes in China based on life cycle analysis. Environ Internat 166:107.

Zhou, D. D., Li, J., Xiong, R. G., Saimaiti, A., Huang, S. Y., Wu, S. X., ... & Li, H. B. (2022). Bioactive compounds, health benefits and food applications of grape. Foods, 11(18), 2755.

INDEX